AF389151

Micro-/Nano-Fiber Sensors and Optical Integration Devices

Micro-/Nano-Fiber Sensors and Optical Integration Devices

Editor

Jin Li

MDPI • Basel • Beijing • Wuhan • Barcelona • Belgrade • Manchester • Tokyo • Cluj • Tianjin

Editor
Jin Li
College of Information
Science and Engineering
Northeastern University
Shenyang
China

Editorial Office
MDPI
St. Alban-Anlage 66
4052 Basel, Switzerland

This is a reprint of articles from the Special Issue published online in the open access journal *Sensors* (ISSN 1424-8220) (available at: www.mdpi.com/journal/sensors/special_issues/Nanofiber_Sensors).

For citation purposes, cite each article independently as indicated on the article page online and as indicated below:

LastName, A.A.; LastName, B.B.; LastName, C.C. Article Title. *Journal Name* **Year**, *Volume Number*, Page Range.

ISBN 978-3-0365-5630-7 (Hbk)
ISBN 978-3-0365-5629-1 (PDF)

Contents

About the Editor

Jin Li

Jin Li is a Research Fellow and Ph.D. supervisor in the College of Information Science and Engineering at Northeastern University, China. He was born in the city of Yuncheng, Shan Xi Province, China, in December 1983. He achieved his Bachelor's Degree in Electronic Science and Technology at Harbin Institute of Technology (HIT), China, in 2007; his Master's Degree in Physical Electronics at HIT, China, in 2009; and his Doctor's Degree in Physical Electronics at HIT, China, in 2013. His research interests include micro/nano-structured photonics, nanomaterials' nonlinear optics, micro/nanofiber sensors, and integrated optical devices. In these areas, he has led more than 10 national and province projects and published more than 50 SCI-indexed papers in the role of first or corresponding author.

Editorial

Micro-/Nano-Fiber Sensors and Optical Integration Devices

Jin Li [1,2]

1   College of Information Science and Engineering, Northeastern University, Shenyang 110819, China;
    lijin@ise.neu.edu.cn
2   Hebei Key Laboratory of Micro-Nano Precision Optical Sensing and Measurement Technology,
    Qinhuangdao 066004, China

Citation: Li, J. Micro-/Nano-Fiber Sensors and Optical Integration Devices. *Sensors* **2022**, *22*, 7673. https://doi.org/10.3390/s22197673

Received: 12 September 2022
Accepted: 23 September 2022
Published: 10 October 2022

Publisher's Note: MDPI stays neutral with regard to jurisdictional claims in published maps and institutional affiliations.

1. High-Performance Micro/Nanofiber Sensors

Because of their strong surface evanescent field, micro-/nanofibers have been used to develop optical sensors and modulation devices with a high performance and integration. In recent years, they have become an important branch of optical fiber optics and novel sensors, and have received extensive attention from researchers from all over the world. On the one hand, the sensing and optical properties of micro/-nanofiber devices can be optimized by introducing different micro/nanostructures through micro/nano-processing technology (femtosecond laser processing, electron beam etching, ion beam engraving, and chemical etching) or through the functionalization of new nanomaterials (film coating and particle doping) in order to achieve high selectivity detection for the desired targets. On the other hand, their tight confinement effect for light based on the high evanescent field coupling efficiency is also conducive to the integration of free-standing optical fibers with typical planar micro-/nanostructures on silicon substrates, making them a promising candidate for exploring nanophotonic integrated devices.

The development of micro/nanofiber sensors and the related integrated systems is a grand project spanning photonics, engineering, and materials science, and will become a hot academic research field. During the development of miniature optical sensors, different materials and micro/nanostructures are reasonably designed and functionalized on ordinary single-mode optical fibers. The combination of various special optical fibers and new micro/nanomaterials has greatly improved the performance of the sensors. In terms of optical integration, micro/nanofiber plays as an independent and movable optical waveguide device and can be conveniently integrated into the two-dimensional chip to realize the efficient transmission and information exchange of optical signals based on optical evanescent field coupling technology. In terms of systematic integration, the unique optical transmission mode of optical fiber shows its great potential in the array and networking of multiple sensor units.

In this book, more than ten research papers were collected and studied on the optical micro/nanofiber devices and related integrated systems, covering the high-performance optical micro/nanofiber sensors, fine characterization technologies for optical micro/nanostructures, weak signal detection technologies in photonic structures, as well as the fiber-assisted highly integrated optical detection systems.

Using the low-cost optical fiber fusion splicers and common single-mode optical fibers, many optical fiber composite structures can be fabricated, such as the fusion joints with offset core, spherical convex cones with enlarging waist, microfiber tapers with the biconical region, microfiber couplers with ultra-thin connecting zone and novel fibers with special materials or structures, which have been designed to develop the compact fiber sensors with high performance.

Deng et al. designed a Mach-Zehnder interferometer using the dislocation fusion technology [1] and modified its surface with 3-aminopropyl-triethoxysilane (APES), which plays as a typical polymer to provide the electrostatic bonding chemical bonds during the layer-by-layer assembly technology for elaborating the optical fibers. The morphological parameters of the APES coating layer were determined and optimized by the microstructural Mach-Zehnder fiber interferometer This technology is crucial for the functional modification of selective enzymes, antibodies, or chemical bonds on the surface of optical fibers. This process is indispensable in the development of fiber biochemical sensors. In this work, the effects of different concentrations, dripping time, and dosage of APES on the transmission spectrum were studied. The best extinction ratio has been experimentally demonstrated to be ~1.165 for the corresponding processing parameters of 2% (concentration), 1 h (immersing time), and 1.5 mL (dosage).

Yi et al. Proposed a convex micro-cone structure for vectorization monitoring of curvature [2]. In this optical fiber structure, the size of the convex cone is very small, which is not different from the ordinary single-mode optical fiber in appearance. A pair of convex micro-cones were used to construct a multimode interferometer. The effects of different overlap lengths and interference region lengths on the interference spectrum were studied to optimize the structural parameters. The curvature responses and dynamic changes in two orthogonal directions have been real-time determined with the sensitivities of $79.1°/m^{-1}$ and $-48.0°/m^{-1}$, corresponding to the resolutions of 2.69×10^{-3} m^{-1} and 4.44×10^{-3} m^{-1}, respectively. The simple and flexible structure, as well as the array and networking advantages of the optical fiber, make this structure promising for real-time monitoring of the direction and extent of curvature changing at multiple points.

Xiang et al. proposed stretchable optical fiber sensors (SPFSs) by encapsulating the dual-cone microfiber in the polydimethylsiloxane (PDMS) membrane [3]. This sensor can be attached to the human skin as a wearable healthcare device to measure the strain, temperature, and humidity caused by pulse, body temperature, and respiration. In this work, the bending and recovering processes have been experimentally monitored to reveal the wrist pulse of ~68 beats per minute. The temperature sensitivity of 0.02 dBm/°C was verified during 30~40 °C when the sensor was fixed on the back of the hand. The proposed sensor was also packaged in a mask to determine the high humidity with a sensitivity of 0.5 dB/%RH.

Gao et al. attached an optical nanofiber coupler to a diaphragm to determine the acoustic in the range of 30~20,000 Hz [4]. The highest sensitivity of 1929 mV/Pa was experimentally demonstrated at the frequency of 120 Hz with a high signal-to-noise ratio of 42.45 dB. The lowest detection limit was verified to be 330 μPa/Hz$^{1/2}$. This acoustic vibration sensor has great application potential in low-frequency vibration ranges, such as seismic wave detection.

The new special optical fiber can be developed to meet some requirements of the optical communication and sensing applications in some extreme environments. Zhang et al. doped different metal oxide materials (Al_2O_3, Y_2O_3, and P_2O_5) into the fiber core to build a specific fiber [5]. They etched the fiber grating structure on the independently designed fiber, and experimentally verified its sensing characteristics to curvature and temperature. The bending sensing sensitivity is 21.85 dB/m^{-1} in the range of 0~1.2 m^{-1}, and the corresponding long-term working fluctuation error is only 0.014 m^{-1}. In the high-temperature range of 20~1000 °C, the temperature response sensitivity is 14.1 pm/°C, and the response time is 0.6 s, when the temperature jumped to 955 °C.

2. Fine Characterization of Micro/Nanostructures

The research and analysis of the morphological parameters and material composition of different micro/nanostructures are also crucial for developing highly integrated sensor devices. The resonance spectrum of optical signals can be generated in the periodic microstructures, which are expected to be applied to the precise identification of biological tissues.

Yang et al. combined ultrasonic (US) and photoacoustic (PA) spectroscopy technology to realize the detection of the periodic microstructure of bone samples [6], where the standing waves were generated among the pores. The US resonance spectra of bone samples after demineralization and decollagenization have been compared and analyzed. It provides a promising technique to early determine osteoporosis.

Brillouin optical time domain analysis (BOTDA) has been widely concerned to monitor the structural health of bridges, dams, tall buildings, equipment, and pipelines. It is worth mentioning that some minor defects can also be detected, as Yuan et al. verified in their work [7], where the micro-cracks with a width of 2~9 μm were accurately measured by combining theoretical models and experiments based on BOTDA technology. Brillouin peak will be significantly increased with the enlarging size of defects. This work proposed an effective method to simulate the strain distribution along the optical fiber due to the micro-cracks.

3. Weak Optical Signal Detection in Photonic Structures

The detection and accurate acquisition of weak optical signals are extremely important for micro/nanoscale photonic devices. Liu et al. designed the cryogenic system with a fluctuation of ± 0.3 °C to smoothly cool down the working temperature of a silicon photomultiplier (SiPM) and compared its dynamic performance in operating voltage and dark noise with those at room temperature [8]. The dark count rate was reduced for 6 orders from 120 kHz/mm^2 to 0.1 Hz/mm^2.

Yang et al. proposed an off-axis Fresnel zone plate based on double exposure points holographic surface interference to eliminate the aberration in recording the instantaneous three-dimensional imaging [9]. The corresponding impact of the introduced method on the Seidel coefficient, average gradient, and signal-to-ratio has been compared. The imaging performance has been significantly improved and verified by many parameters, such as spherical aberration W040, coma aberration W131, image dispersion W222, field curvature W220, average gradient, and signal-to-noise ratio.

4. High Integration Optical Fiber Assisted-Sensing System

The optical fiber can be conveniently connected to the high-performance optical system to achieve the efficient transmission and collection of optical signals, so as to improve the integration density of laser detection devices.

Hu et al. designed an optical fiber integrated dissolved gas detector to real-time monitor the dissolved carbon dioxide in seawater based on ring cavity down spectroscopy [10]. The sensor effectively improves the water gas separation efficiency, and the detection rate is 10 times higher than that of the ordinary sensing method. The practical test of offshore seawater shows the practicability of the proposed sensor system, where real-time monitoring was achieved for the water-soluble carbon dioxide with a concentration of ~950 ppm. The sensor system is highly adaptable to the complex environment of the deep sea (the depth of <4500 m and low temperature) for a long time.

Zhou et al. used the Sagnac microfiber loop to measure the concentration of airborne molecular contaminants (AMCs) in the air environment, which becomes a serious threat to the service life of UV high-energy lasers [11]. The mesoporous silica was elaborated on the surface of the Sagnac ring by microheater brushing method and dip coating technique. It can absorb the AMCs to cause a change in its effective refractive index. The sensitivity was determined to be 0.11 nm (mg/m^3). This work supplies a promising way to detect trace contaminates in the air environment around the lens or micro/nanostructures.

Funding: This research was funded by the National Key R&D Program of China (2019YFB2006001), and Hebei Natural Science Foundation (F2020501040).

Conflicts of Interest: The authors declare no conflict of interest.

References

1. Deng, H.; Chen, X.; Huang, Z.; Kang, S.; Zhang, W.; Li, H.; Shu, F.; Lang, T.; Zhao, C.; Shen, C. Optical Fiber Based Mach-Zehnder Interferometer for APES Detection. *Sensors* **2021**, *21*, 5870. [CrossRef] [PubMed]
2. Yi, D.; Wang, L.; Geng, Y.; Du, Y.; Li, X.; Hong, X. Multiplexed Weak Waist-Enlarged Fiber Taper Curvature Sensor and Its Rapid Inline Fabrication. *Sensors* **2021**, *21*, 6782. [CrossRef] [PubMed]
3. Xiang, S.; You, H.; Miao, X.; Niu, L.; Yao, C.; Jiang, Y.; Zhou, G. An Ultra-Sensitive Multi-Functional Optical Micro/Nanofiber Based on Stretchable Encapsulation. *Sensors* **2021**, *21*, 7437. [CrossRef] [PubMed]
4. Gao, X.; Wen, J.; Wang, J.; Li, K. Broadband Acoustic Sensing with Optical Nanofiber Couplers Working at the Dispersion Turning Point. *Sensors* **2022**, *22*, 4940. [CrossRef] [PubMed]
5. Zhang, Y.; Zhang, Y.; Hu, X.; Wu, D.; Fan, L.; Wang, Z.; Kong, L. An Ultra-High-Resolution Bending Temperature Decoupled Measurement Sensor Based on a Novel Core Refractive Index-Like Linear Distribution Doped Fiber. *Sensors* **2022**, *22*, 3007. [CrossRef] [PubMed]
6. Yang, L.; Chen, C.; Zhang, Z.; Wei, X. Diagnosis of Bone Mineral Density Based on Backscattering Resonance Phenomenon Using Coregistered Functional Laser Photoacoustic and Ultrasonic Probes. *Sensors* **2021**, *21*, 8243. [CrossRef] [PubMed]
7. Yuan, B.; Ying, Y.; Morgese, M.; Ansari, F. Theoretical and Experimental Studies of Micro-Surface Crack Detections Based on BOTDA. *Sensors* **2022**, *22*, 3529. [CrossRef] [PubMed]
8. Liu, F.; Fan, X.; Sun, X.; Liu, B.; Li, J.; Deng, Y.; Jiang, H.; Jiang, T.; Yan, P. Characterization of a Mass-Produced SiPM at Liquid Nitrogen Temperature for CsI Neutrino Coherent Detectors. *Sensors* **2022**, *22*, 1099. [CrossRef]
9. Yang, L.; Ma, Z.; Liu, S.; Jiao, Q.; Zhang, J.; Zhang, W.; Pei, J.; Li, H.; Li, Y.; Zou, Y.; et al. Study of the Off-Axis Fresnel Zone Plate of a Microscopic Tomographic Aberration. *Sensors* **2022**, *22*, 1113. [CrossRef]
10. Hu, M.; Chen, B.; Yao, L.; Yang, C.; Chen, X.; Kan, R. A FiberIntegrated CRDS Sensor for In-Situ Measurement of Dissolved Carbon Dioxide in Seawater. *Sensors* **2021**, *21*, 6436. [CrossRef] [PubMed]
11. Zhou, G.; Xiang, S.; You, H.; Li, C.; Niu, L.; Jiang, Y.; Miao, X.; Xie, X. A Novel Airborne Molecular Contaminants Sensor Based on Sagnac Microfiber Structure. *Sensors* **2022**, *22*, 1520. [CrossRef]

Communication

Optical Fiber Based Mach-Zehnder Interferometer for APES Detection

Huitong Deng, Xiaoman Chen, Zhenlin Huang, Shiqi Kang, Weijia Zhang, Hongliang Li, Fangzhou Shu, Tingting Lang, Chunliu Zhao and Changyu Shen *

Institute of Optoelectronic Technology, China Jiliang University, Hangzhou 310018, China;
1700201401@cjlu.edu.cn (H.D.); p1904085202@cjlu.edu.cn (X.C.); p20040854033@cjlu.edu.cn (Z.H.);
11801400310@cjlu.edu.cn (S.K.); 1700603202@cjlu.edu.cn (W.Z.); lihongliang@cjlu.edu.cn (H.L.);
shufangzhou@cjlu.edu.cn (F.S.); langtingting@cjlu.edu.cn (T.L.); zhaochunliu@cjlu.edu.cn (C.Z.)
* Correspondence: shenchangyu@cjlu.edu.cn

Abstract: A 3-aminopropyl-triethoxysilane (APES) fiber-optic sensor based on a Mach–Zehnder interferometer (MZI) was demonstrated. The MZI was constructed with a core-offset fusion single mode fiber (SMF) structure with a length of 3.0 cm. As APES gradually attaches to the MZI, the external environment of the MZI changes, which in turn causes change in the MZI's interference. That is the reason why we can obtain the relationships between the APES amount and resonance dip wavelength by measuring the transmission variations of the resonant dip wavelength of the MZI. The optimized amount of 1% APES for 3.0 cm MZI biosensors was 3 mL, whereas the optimized amount of 2% APES was 1.5 mL.

Keywords: optical fiber biosensor; Mach-Zehnder Interferometer; APES detection; condensation reaction

Citation: Deng, H.; Chen, X.; Huang, Z.; Kang, S.; Zhang, W.; Li, H.; Shu, F.; Lang, T.; Zhao, C.; Shen, C. Optical Fiber Based Mach-Zehnder Interferometer for APES Detection. *Sensors* **2021**, *21*, 5870. https://doi.org/10.3390/s21175870

Academic Editor: Jin Li

Received: 4 August 2021
Accepted: 26 August 2021
Published: 31 August 2021

Publisher's Note: MDPI stays neutral with regard to jurisdictional claims in published maps and institutional affiliations.

1. Introduction

3-aminopropyl-triethoxysilane (APES) is widely used in combination with crosslinkers, owing to its reactivity with different functional groups, such as aldehyde, carboxylic acid, and epoxy [1,2]. In fiber-optic nucleic acid biosensors, the silanization process is the chemical reaction between APES and a hydroxyl group, which will form a covalent linkage [3–7]. Additionally, the covalent linkage can fix amino groups on the surface of the fiber-optic Mach–Zehnder interferometer (MZI). Because of the considerable stability of the covalent linkage, optical fiber modified with APES has found widespread application [8]. Specifically, APES is a type of coupling reagent that can link two functional groups together, such as hydroxy and aldehyde [9].

While the condensation reaction between APES and hydroxyl group is occurring on the surface of the MZI, APES will also react with water [10]. To prevent APES from reacting with water, toluene is usually used as the solvent to prepare the APES solution. However, when APES is used in experiments, especially in fiber-optic biosensor experiments, there is no clear rule or experience to determine the amounts and concentrations of the APES solution. Unsuitable amounts and concentrations often lead to failure of the experiments. Usually, in APES-based fiber-optic sensing experiments, the APES concentrations mostly range from 1% to 2%, and the immersion time ranges from 45 min to 2 h [11–17]. Therefore, it is valuable to determine suitable APES amounts and concentrations for the fabrication of optical fibers coated with well-proportioned APES films.

In this paper, an optical fiber-based MZI for APES detection was proposed. The different amounts and concentrations of APES on the fiber surface can be monitored by the MZI's interference patterns. The relationship between the amount of APES and the resonance dip wavelength was obtained. Optimized amounts of 1% APES and 2% APES for fiber-optic biosensors were also obtained.

2. Experimental Methods and Principle

2.1. Sensor Fabrication and Principle

A schematic diagram of the APES fiber-optic sensor based on MZI is shown in Figure 1. A fiber-coupled broadband source (BBS) with the wavelength range of 1500–1620 nm is used as the input light source. The transmission spectrum of the MZI is recorded by an optical spectrum analyzer (OSA: AQ6317B, YOKOGAWA, Japan) with a wavelength resolution of 0.02 nm. The BBS is coupled into the single-mode fiber (SMF), connecting to the OSA.

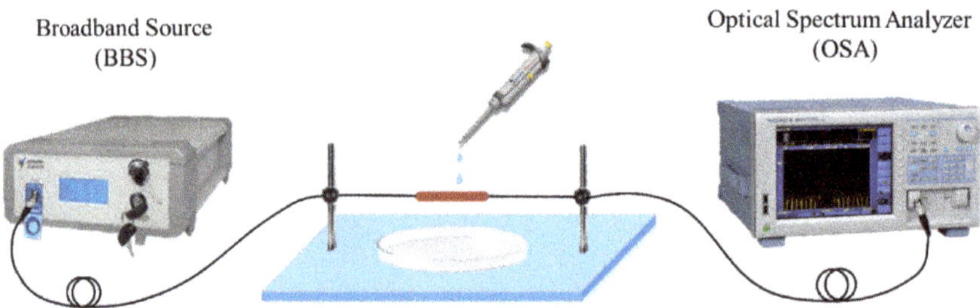

Figure 1. Schematic diagram of the APES fiber-optic sensor.

In the paper, a developed core-offset fusion SMF structure was used for the advantages of low cost and easy operation. Figure 2 shows an MZI based on the core-offset fusion SMF structure, where Figure 2a is a partially enlarged drawing of the MZI, and Figure 2b shows the core-offset connect joints on the fusion splicer screen. The fiber used in the experiment was Corning® SMF-28e+® BB. The core and cladding diameters of the SMF were 9 μm and 125 μm, respectively. The core-offset connect joints showing on the fusion splicer screen had no built-in rule scale. So, the cladding diameter has been marked as 125 um in Figure 2b.

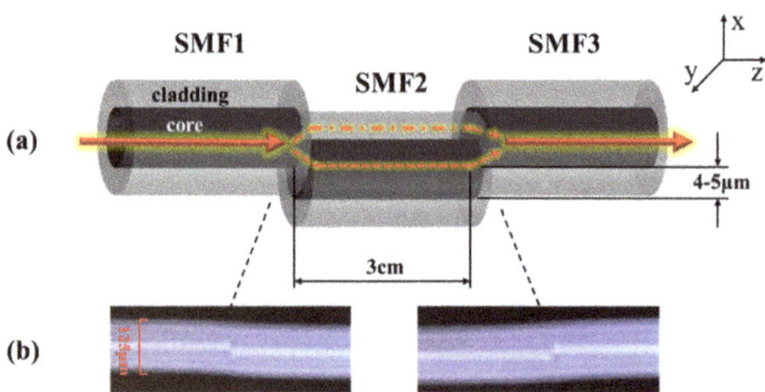

Figure 2. (**a**) The partially enlarged drawing of the MZI (**b**) Pictures of the core-offset connect joints showing on the fusion splicer screen.

The core-offset structure was formed by using a commercial electric-arc fusion splicer (Fujikura FSM-60s). Along the x axis direction of SMF1, SMF2 was shifted downward by 4 to 5 μm with a length of 3.0 cm. SMF2 and SMF3 also adopted core-offset fusion. SMF3 moved upward by 4 to 5 μm along the x axis direction of SMF2 [18].

For the core-offset structure, the light from SMF1 cladding is partly coupled to the SMF2 core to excite the core-mode light, and the rest of the light from SMF1 cladding enters the SMF2 cladding to excite the cladding-mode light. After propagating through SMF2, the component of core-mode light of SMF2 and component of cladding-mode light of SMF2 couple to the SMF3 core together, forming a type of MZI [19].

In the MZI, the phase difference $\Delta\phi$ between the lights propagating in SMF2 cladding and core can be expressed as [20],

$$\Delta\phi = 2\pi(n_{eff}^{co} - n_{eff}^{cl})L/\lambda \tag{1}$$

where n_{eff}^{co} and n_{eff}^{cl} respectively represent the effective refractive indices of core-mode and cladding-mode, λ is the center wavelength of the input light, and L is the effective length of the MZI. The intensities I of the interference light wave of the MZI can be expressed as,

$$I = I_1 + I_2 + 2\sqrt{I_1 I_2}\cos(\Delta\phi) \tag{2}$$

where I_1 and I_2 represent the intensities of the cladding-mode light and the core-mode light, respectively. Additionally, the fringe visibility K of the interference pattern is,

$$K = \frac{2\sqrt{I_1 I_2}}{I_1 + I_2} \tag{3}$$

It can be seen from Equation (3) that the smaller the difference between I_1 and I_2, the larger K, and consequently the interference peak becomes stronger. In the output interferogram, the resonant dip wavelength corresponding to the interference peak can be expressed as,

$$\lambda_r = 2(n_{eff}^{co} - n_{eff}^{cl})L/2k + 1 \tag{4}$$

where λ_r represents the resonant dip wavelength and k represents the wave number. When the external environment changes, the interference peak changes significantly.

2.2. APES Film Fabrication

Ninety-eight percent 3-aminopropyl-triethoxysilane (APES) (China Beijing Solibao Technology Co., Ltd.) was stored at room temperature (RT). Ninety-eight percent toluene and KOH used in this study were of analytical grade.

First, the MZI was rinsed 3 times with absolute ethanol and deionized water, respectively, to eliminate residue and dirt on the surface of the optical fiber, then dried naturally. Second, it was immersed in 0.1 M KOH for 1 h to carry out hydroxylation and form a layer of film with hydroxylated group as shown in Figure 3a. Next, after generating the hydroxylated group, the MZI was silanized with APES to form a layer of film with amino group; the reaction schematic diagram is shown in Figure 3b. In the experiments, 1% APES and 2% APES were adopted to drip onto the optical fiber surface, respectively. The amount of both APES concentrations ranged from 0 to 7.5 mL in 1.5 mL steps. By measuring the transmission intensities of the resonant dip wavelengths of the MZI, the relationships between the APES concentrations and resonance dip wavelength transmission intensities were obtained.

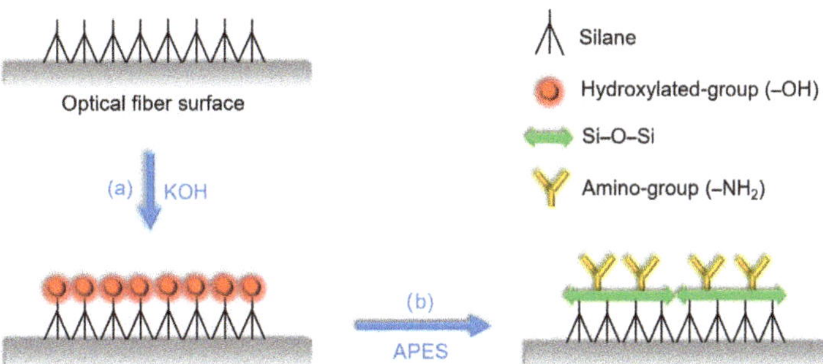

Figure 3. Schematic diagram of (**a**) the hydroxylation process by KOH (**b**) the silanization process by APES.

3. Results and Discussion

3.1. Validation of the Dripping Method's Feasibility

In order to verify that adhesion between the optical fiber and APES could be achieved in a short time, we recorded and analyzed the transmission spectra of MZI immersed in 1% APES toluene solution for 1 h. Figure 4a shows the transmission spectra of the MZI corresponding to the MZI immersion time with 1% APES, and Figure 4b shows the relationship between the MZI immersion time with 1% APES and the resonant wavelength's transmission intensities. When the MZI was immersed in 1% APES, the transmission intensities decreased instantly due to the condensation reaction. There was no drift for 1 h following, which illustrated that adhesion between the optical fiber and APES could be achieved in a short time. Removing the MZI from the APES 1 h later, the transmission intensities increased owing to the change in external environment from liquid to air.

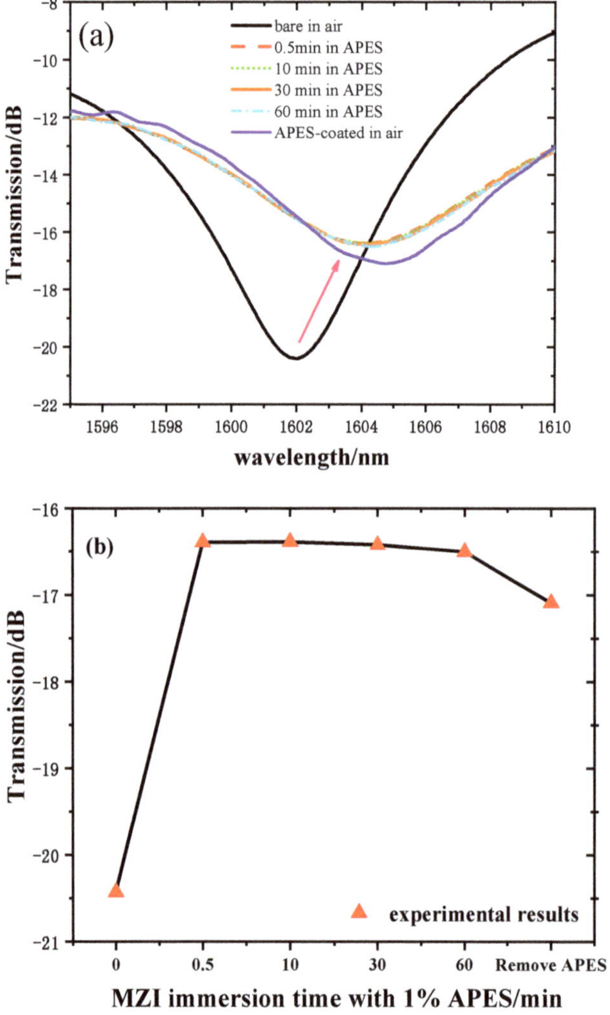

Figure 4. (**a**) Transmission spectra of MZI corresponding to immersion time in 1% APES. (**b**) The relationship between MZI immersion time in 1% APES and transmission intensities.

3.2. Validation of the Necessity of Hydroxylation

A comparative experiment was performed to verify that APES cannot be directly combined with bare optical fiber, but can be combined with hydroxyl-modified optical fiber. After cleaning the bare optical fiber with absolute ethanol and deionized water, 1% APES was directly dripped onto the MZI without hydroxyl group. Cleaning the MZI with deionized water again, the transmission spectra underwent a change consistent with that of the bare MZI in air, as shown in the Figure 5.

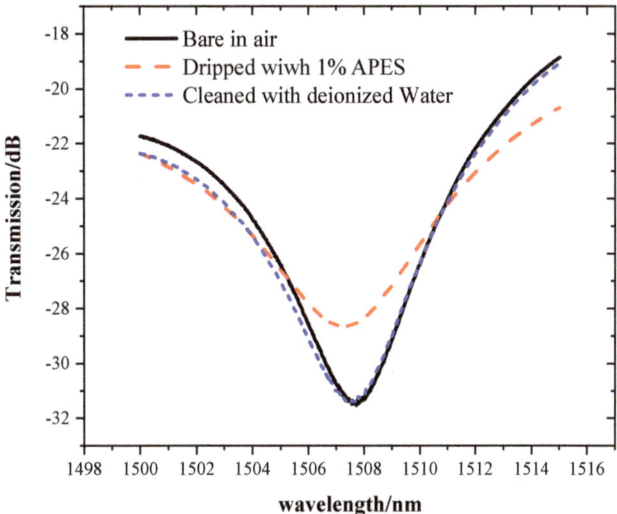

Figure 5. The transmission spectra of the MZI without hydroxyl group.

This phenomenon showed that if the fiber is not modified with hydroxyl, APES cannot be directly combined with the bare optical fiber. The spectra changed after adding APES. While cleaning the MZI with deionized water, the transmission spectra underwent a change consistent with that of the bare MZI in air, which meant that APES adsorbed onto a fiber surface without hydroxyl could be removed.

3.3. Determination of APES Amount

The refractive indices of 1% APES toluene solution and fiber cladding are 1.496 and 1.462, respectively. Therefore, when dripping 1% APES onto the optical fiber surface, the effective refractive index of the cladding mode n_{eff}^{cl} increased. According to Equations (1) and (2), the phase difference $\Delta\varphi$ decreased, while the interference light intensity I increased. Concerning the resonant wavelength of the MZI, when n_{eff}^{cl} increases, the resonant wavelength λ_r should decrease. Additionally, the resonant wavelength of the MZI should shift to the short wavelength. However, because the function of APES is adhesion, stress may be generated on the MZI surface, affecting the resonant wavelength's drift.

Figure 6a shows the transmission spectra of 1% APES amounts from 0 to 7.5 mL. The response time was very fast; the spectrum drifted immediately after drying the APES, and it stabilized within 10 min. Because there are many interference resonant wavelengths in the Mach–Zehnder interference spectrum, one resonant wavelength (near 1610 nm for instance) was selected for sensing. The transmission intensity of the resonant wavelength of 1612 nm was chosen as the sample wavelength. Figure 6b shows the relationship between the 1% APES amount and the resonant wavelength's extinction ratio. Obviously, there was no linear relationship between the amounts of APES and the extinction ratio at the resonant wavelength of 1602 nm. When the applied amount of APES was about 1.5 mL, the hydroxylated group on the optical fiber was almost saturated. Therefore, continued increases in the amount of APES did not result in significant fluctuations in transmission intensity compared to that obtained with 1.5 mL. Due to the gradual saturation of the condensation reaction, the transmission intensities increased significantly and tended to stabilize after dripping 3.0 mL of 1% APES, which illustrated that the optimal amount of 1% APES was 3.0 mL.

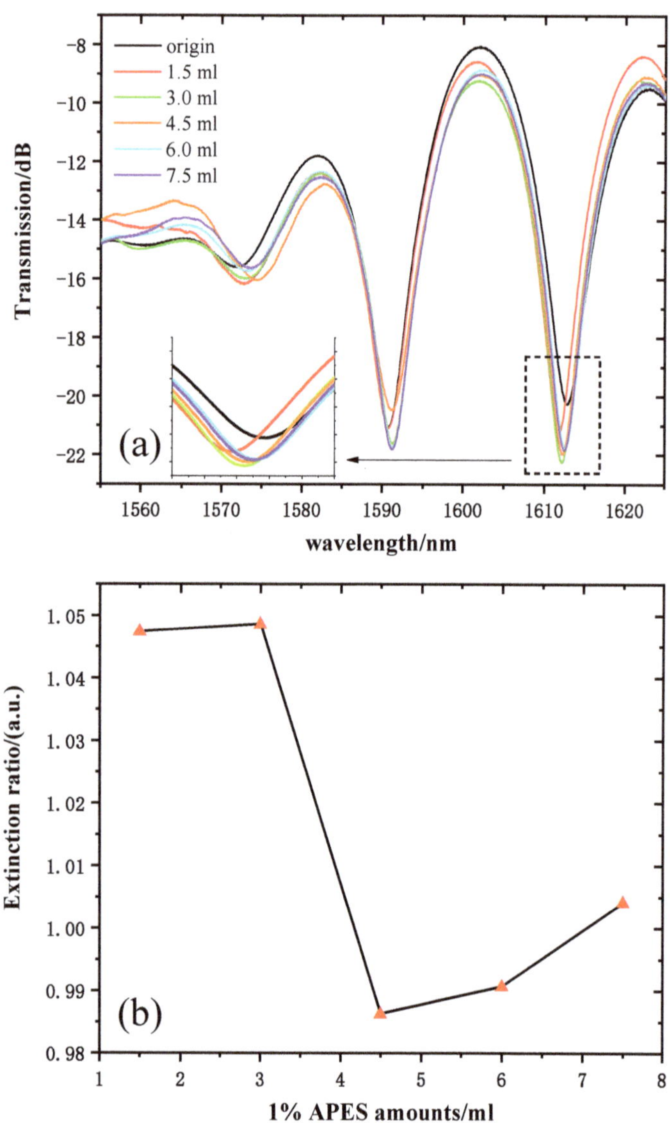

Figure 6. (a) MZI transmission spectra corresponding to the MZI dripped with 1% APES. (b) The relationship between the amount of 1% APES and the extinction ratio.

The refractive indices of 2% APES toluene solution and fiber cladding are 1.493 and 1.462, respectively. When 1.5 mL 2% APES was dripped onto the MZI surface, the effective refractive index of cladding mode n_{eff}^{cl} also increased. Thus, the same principle could illustrate that the transmission intensities increased.

The transmission spectra (near 1600 nm, for instance) of 2% APES amounts ranging from 0 to 7.5 mL are shown in Figure 7a. The transmission intensity of the resonant wavelength of 1612 nm was chosen to reflect the variations. Figure 7b shows the relationship between the amounts of 2% APES and resonant wavelength transmission intensities. Due to the gradual saturation of the condensation reaction, the transmission intensities increased significantly and tended to stabilize after dripping 1.5 mL of 2% APES, which illustrated that the optimal amount of 2% APES was 1.5 mL.

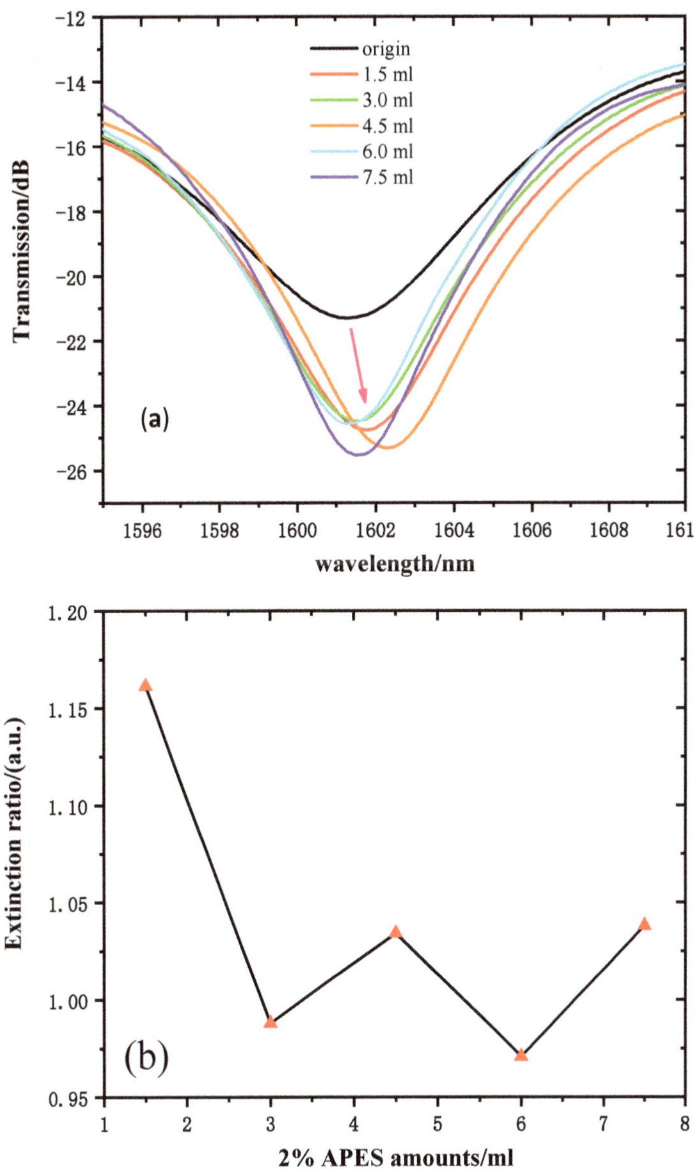

Figure 7. (**a**) Transmission spectra of MZI with fiber dripped with 2% APES. (**b**) The relationship between the amount of 2% APES and the extinction ratio.

3.4. SEM of the APES-Modified Optical Fiber Surface

The 1% APES film on the optical fiber surface was examined using scanning electron microscopy (SEM) to verify that adhesion between the optical fiber and APES could be achieved with the dripping method. Figure 8a is the 500 times magnified side view of the APES film after soaking the optical fiber in 1% APES for 1 h, and Figure 8b is the 500 times magnified side view of the APES film generated by dripping with 3.0 mL 1% APES onto the optical fiber surface. Compared to the soaking method, the APES film generated by the dripping method was more uniform and thinner, which was conducive to the detection of other reagents in the later steps.

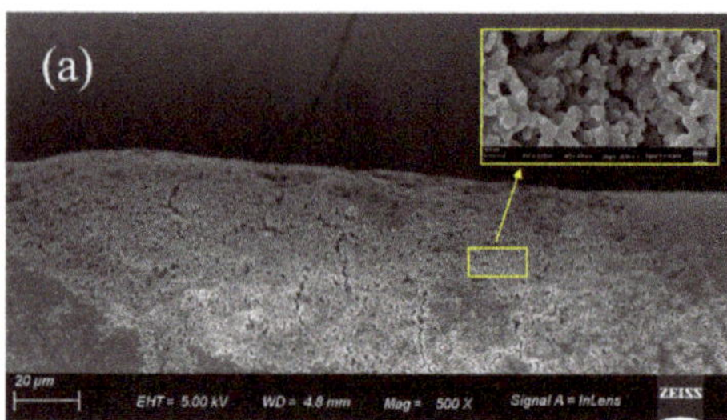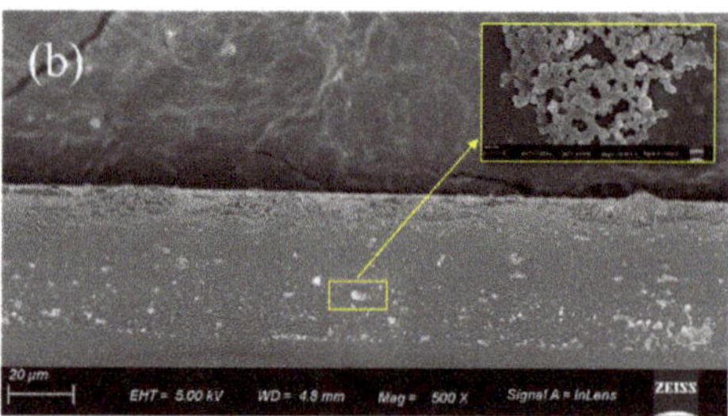

Figure 8. SEM micrographs detailing morphologies of APES film on the optical fiber surface: (**a**) soaked with 1% APES for 1 h; (**b**) dripped with 3 mL 1% APES. Inset: 5k times magnified view of the APES film.

4. Conclusions

In summary, an optical fiber-based MZI for APES detection was proposed. When APES was dripped onto the hydroxyl-modified MZI, the condensation reaction affected the effective refractive index of the cladding mode, which in turn caused a change in the resonant wavelength's transmission intensity. Measuring the transmission intensities of the MZI revealed that increasing the amount of APES caused the transmission intensities to increase at the beginning and tend toward stabilization with the gradual saturation of the condensation reaction. The optimal amount of 1% APES for fiber-optic biosensors was 3.0 mL, while the optimal amount of 2% APES was 1.5 mL. Given its advantage of easy fabrication, the core-offset-based MZI structure has potential applications for many types of fiber-optic sensors. In addition, in the field of fiber-optic sensing, APES-modified fiber-optic biosensors with amino groups can be widely used in combination with different functional groups for the immobilization of proteins or nucleic acids.

Author Contributions: Conceptualization, C.S., H.D., X.C.; methodology, W.Z.; investigation, S.K.; data curation, Z.H.; writing—original draft preparation, H.D.; writing—review and editing, C.S.; H.L., F.S., T.L. and C.Z. All authors have read and agreed to the published version of the manuscript.

Funding: Zhejiang Provincial Natural Science Foundation of China (LY21F050006), National Natural Science Foundation of China (11874332, 61727816, 12004362, 61875251, 61775202), and Key R & D plan of Zhejiang Province (2021C01179).

Institutional Review Board Statement: Not applicable.

Informed Consent Statement: Not applicable.

Conflicts of Interest: The authors declare no conflict of interest.

References

1. Howarter, J.A.; Youngblood, J.P. Optimization of silica silanization by 3-aminopropyltriethoxysilane. *Langmuir ACS J. Surf. Colloids* **2006**, *22*, 11142–11147. [CrossRef] [PubMed]
2. Zhu, C.; Li, S.; Cong, X.; Rudd, C.; Liu, X. The effect of 3-aminopropyltriethoxysilanetreatment on recycled carbon fibers reinforced composites. In Proceedings of the 21st National Conference on Composite Materials (NCCM-21), Inner Mongolia, China, 29 July–1 August 2020.

3. Bañuls, M.-J.; Puchades, R.; Maquieira, Á. Chemical surface modifications for the development of silicon-based label-free integrated optical (IO) biosensors: A review. *Anal. Chim. Acta* **2013**, *777*, 1–16. [CrossRef] [PubMed]

4. Huertas, C.S.; Calvo-Lozano, O.; Mitchell, A.; Lechuga, L.M. Advanced Evanescent-Wave Optical Biosensors for the Detection of Nucleic Acids: An Analytic Perspective. *Front. Chem.* **2019**, *7*, 724. [CrossRef]

5. Kaur, A.; Kapoor, S.; Bharti, A.; Rana, S.; Chaudhary, G.R.; Prabhakar, N. Gold-platinum bimetallic nanoparticles coated 3-(aminopropyl)triethoxysilane (APTES) based electrochemical immunosensor for vitamin D estimation. *J. Electroanal. Chem.* **2020**, *873*, 114400. [CrossRef]

6. Bharti, A.; Mittal, S.; Rana, S.; Dahiya, D.; Agnihotri, N.; Prabhakar, N. Electrochemical biosensor for miRNA-21 based on gold-platinum bimetallic nanoparticles coated 3-aminopropyltriethoxy silane. *Anal. Biochem.* **2020**, *609*, 113908. [CrossRef] [PubMed]

7. Xu, X.; Song, X.; Nie, R.; Yang, Y.; Chen, Y.; Yang, L. Ultra-sensitive capillary immunosensor combining porous-layer surface modification and biotin-streptavidin nano-complex signal amplification: Application for sensing of procalcitonin in serum. *Talanta* **2019**, *205*, 120089. [CrossRef] [PubMed]

8. Rao, X.; Tatoulian, M.; Guyon, C.; Ognier, S.; Chu, C.; Hassan, A.A. A Comparison Study of Functional Groups (Amine vs. Thiol) for Immobilizing AuNPs on Zeolite Surface. *Nanomaterials* **2019**, *9*, 1034. [CrossRef] [PubMed]

9. Lee, M.R.; Shin, I. Fabrication of chemical microarrays by efficient immobilization of hydrazide-linked substances on epoxide-coated glass surfaces. *Angew. Chem.* **2005**, *44*, 2881. [CrossRef] [PubMed]

10. Han, S. *Research on Hg^{2+} and Pb^{2+} Evanescent Wave Biosensing Detection Technology Based on Functional Nucleic Acid*; Tsinghua University: Beijing, China, 2016.

11. Han, S.; Zhou, X.; Tang, Y.; He, M.; Zhang, X.; Shi, H.; Xiang, Y. Practical, highly sensitive, and regenerable evanescent-wave biosensor for detection of Hg^{2+} and Pb^{2+} in water. *Biosens. Bioelectron.* **2016**, *80*, 265–272. [CrossRef] [PubMed]

12. Ainuddin, N.H.; Chee, H.Y.; Ahmad, M.Z.; Mahdi, M.A.; Abu Bakar, M.H.; Yaacob, M.H. Sensitive Leptospira DNA detection using tapered optical fiber sensor. *J. Biophotonics* **2018**, *11*, e201700363. [CrossRef] [PubMed]

13. Zhang, G.J.; Luo, Z.H.H.; Huang, M.J.; Tay, G.K.I.; Lim, E.J.A. Morpholino-functionalized silicon nanowire biosensor for sequence-specific label-free detection of DNA. *Biosens. Bioelectron.* **2010**, *25*, 2447–2453. [CrossRef]

14. Rong, G.; Najmaie, A.; Sipe, J.E.; Weiss, S.M. Nanoscale porous silicon waveguide for label-free DNA sensing. *Biosens. Bioelectron.* **2008**, *23*, 1572–1576. [CrossRef] [PubMed]

15. De Tommasi, E.; De Stefano, L.; Rea, I.; Di Sarno, V.; Rotiroti, L.; Arcari, P.; Lamberti, A.; Sanges, C.; Rendina, I. Porous Silicon Based Resonant Mirrors for Biochemical Sensing. *Sensors* **2008**, *8*, 6549–6556. [CrossRef]

16. Massad-Ivanir, N.; Shtenberg, G.; Zeidman, T.; Segal, E. Construction and Characterization of Porous SiO$_2$/Hydrogel Hybrids as Optical Biosensors for Rapid Detection of Bacteria. *Adv. Funct. Mater.* **2010**, *20*, 2269–2277. [CrossRef]

17. Grego, S.; McDaniel, J.R.; Stoner, B.R. Wavelength interrogation of grating-based optical biosensors in the input coupler configuration. *Sens. Actuators B Chem.* **2008**, *131*, 347–355. [CrossRef]

18. Gong, J.; Shen, C.; Xiao, Y.; Liu, S.; Zhang, C.; Ding, Z.; Deng, H.; Fang, J.; Lang, T.; Zhao, C.; et al. High sensitivity fiber temperature sensor based PDMS film on Mach-Zehnder interferometer. *Opt. Fiber Technol.* **2019**, *53*, 102029. [CrossRef]

19. Yu, F.; Liu, B.; Zou, Q.; Xue, P.; Zheng, J. Influence of temperature on the refractive index sensor based on a core-offset in-line fiber Mach-Zehnder interferometer. *Opt. Fiber Technol.* **2020**, *58*, 102293. [CrossRef]

20. Zaini, M.K.A.; Lee, Y.-S.; Ismail, N.; Lim, K.-S.; Udos, W.; Zohari, M.H.; Yang, H.-Z.; Ahmad, H. In-fiber Fabry Perot interferometer with narrow interference fringes for enhanced sensitivity in elastic wave detection. *Opt. Fiber Technol.* **2019**, *53*, 102021. [CrossRef]

Communication

Multiplexed Weak Waist-Enlarged Fiber Taper Curvature Sensor and Its Rapid Inline Fabrication

Duo Yi [1], Lina Wang [1], Youfu Geng [1,*], Yu Du [1], Xuejin Li [2] and Xueming Hong [1]

[1] College of Physics and Optoelectronic Engineering, Shenzhen University, Shenzhen 518061, China; yiduo@szu.edu.cn (D.Y.); wanglina2017@email.szu.edu.cn (L.W.); duyu@szu.edu.cn (Y.D.); xmhong@szu.edu.cn (X.H.)
[2] School of Science and Engineering, Chinese University of Hong Kong, Shenzhen 518172, China; lixuejin@szu.edu.cn
* Correspondence: gengyf@szu.edu.cn

Abstract: This study proposes a multiplexed weak waist-enlarged fiber taper (WWFT) curvature sensor and its rapid fabrication method. Compared with other types of fiber taper, the proposed WWFT has no difference in appearance with the single mode fiber and has ultralow insertion loss. The fabrication of WWFT also does not need the repeated cleaving and splicing process, and thereby could be rapidly embedded into the inline sensing fiber without splicing point, which greatly enhances the sensor solidity. Owing to the ultralow insertion loss (as low as 0.15 dB), the WWFT-based interferometer is further used for multiplexed curvature sensing. The results show that the different curvatures can be individually detected by the multiplexed interferometers. Furthermore, it also shows that diverse responses for the curvature changes exist in two orthogonal directions, and the corresponding sensitivities are determined to be $79.1°/m^{-1}$ and $-48.0°/m^{-1}$ respectively. This feature can be potentially applied for vector curvature sensing.

Keywords: curvature sensor; fiber interferometer; waist-enlarged fiber taper

Citation: Yi, D.; Wang, L.; Geng, Y.; Du, Y.; Li, X.; Hong, X. Multiplexed Weak Waist-Enlarged Fiber Taper Curvature Sensor and Its Rapid Inline Fabrication. *Sensors* **2021**, *21*, 6782. https://doi.org/10.3390/s21206782

Academic Editor: Jin Li

Received: 15 September 2021
Accepted: 11 October 2021
Published: 13 October 2021

Publisher's Note: MDPI stays neutral with regard to jurisdictional claims in published maps and institutional affiliations.

1. Introduction

Curvature sensing is essential in various fields such as smart robot [1], engineering structure [2] and medical image [3], etc. When compared with the traditional resistive curvature sensors, the optical fiber curvature sensor owns the advantages such as flexibility, small size, low cost, immunity to electromagnetic interference and easiness to form sensing network [4,5]. Nowadays, optical fiber curvature sensors based on intermodal interferometer have been widely reported [6–8]. Particularly in recent years, the use of the microstructured fiber as the transducing platform promotes the exploitation of various curvature sensors based on photonic crystal fiber [9], hollow-core fiber [7], few-mode fiber [10], four-core fiber [11] or Fabry-Perot [12], etc. However, for these intermodal interferometers, post-processing techniques such as fiber tapering [13], side-polishing [14] and cladding etching [15] are always used to modify the fiber structure and convert part of fundamental mode into the high-order mode in order to construct an optical arm. It inevitably results in a large insertion loss and structure fragileness, which makes the essentially multiplexed curvature sensing difficult to be achieved with these techniques.

In our previous work [16], a waist-enlarged fiber taper sensor has been proposed for high-temperature sensing. It has a remarkable and distinguishable fiber structure, a strongly waist-enlarged taper, to excite high-order fiber modes. This type of fiber taper can be named as a strong waist-enlarged fiber taper (SWFT) and has been comprehensively used to construct fiber sensors for the sensing applications such as temperature [16], strain [17], refractive index [18,19] and humidity [20]. It owns the advantages of robust structure, easy fabrication and flexible mode excitation, etc. However, it needs repeated cleaving and splicing during the fabrication process, and the structural health of the sensor still needs to be improved due to the embedded splicing points in the fiber. Besides, similar to the intermodal interferometers mentioned in the previous paragraph, the fabricated SWFT also faces the problem of large transmission loss, which results in a poor signal-noise ratio, but in contrast, presents a moderate spectral dynamic range over 10 dB [16].

In this study, a weak waist-enlarged fiber taper (WWFT) and its rapid inline fabrication method are further proposed, and the application for multiplexed curvature measurement is studied. The WWFT could be fabricated rapidly with a weak electric arc discharge released on the inline sensing fiber, and the repeated cleaving-splicing process is no longer needed. It has no fusion points and no obvious structural deformation in appearance compared with general single mode fiber (SMF). Therefore, the insertion loss of the WWFT-based interferometer is greatly reduced, which makes it possible to realize multiplex curvature sensing with phase-demodulation scheme. Unfortunately, the WWFT device presents a relatively low spectral dynamic range. Finally, the experimental results of curvature measurement demonstrate that the proposed sensor has diverse responses for the curvature changes in two orthogonal directions, showing its great potential applications in vector curvature sensing.

2. Sensor Fabrication

Figure 1a shows the schematic of WWFT and the constructed intermodal interferometer, which has no difference in appearance with the SMF. Figure 1b shows the specific inline fabrication process of WWFT. It is fabricated by using the manual splicing mode of the fiber splicer (Fujikura Ltd., FSM 60S, Tokyo, Japan) in our experiment. Firstly, two segments of SMFs with polymer coating stripped and end cleaved are prepared, and they are separately put onto the motor stages of the fiber holder. The motors are impelled and the fiber end faces are aligned, then the splicing process is temporarily suspended. In this step, the motors are adjusted to avoid fiber bending during arc discharge. Secondly, an SMF whose polymer coating has been stripped is used to replace the aligned fibers on the motor stages, and arc discharge is then released when the two motor stages are pushed forward with a designed overlap distance, defined as L_{vp}. In this study, three L_{vp} values of 10, 20 and 30 µm are selected to fabricate the WWFT in the experiment for comparison. Using this method, the WWFT is rapidly embedded into the sensing SMF fiber. The arc discharge intensity and duration time are set as 'standard mode' and 200 ms respectively. Note that, if the motor can be automatically aligned, the aligned process with two cleaved fibers in the first step is no longer needed and the duration of fabrication process could be shortened to just a few seconds. Therefore, the WWFT is possible to be massively fabricated by a specially designed arc releasing electrical and mechanical structure for rapidly arranging a large-scale sensing network. Figure 1c shows the side views of the WWFT. It can be seen that, different from the SWFT, it has no difference in appearance with SMF, and no splicing point is embedded in the sensing fiber, which gives a strong physical structure and ultralow insertion.

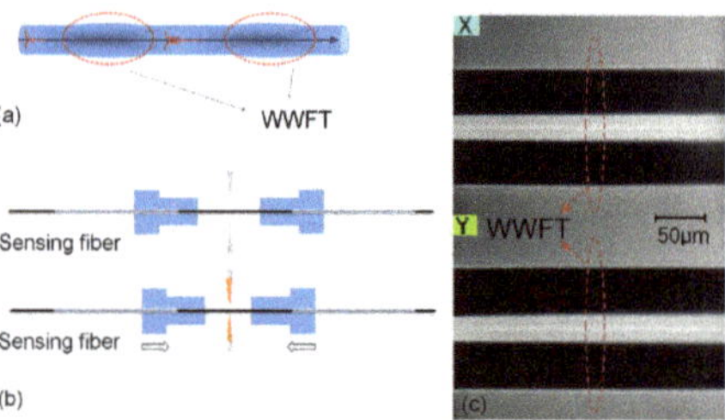

Figure 1. Schematic of WWFT and its fabrication: (**a**) Schematic of WWFT and the constructed intermodal interferometer; (**b**) Inline fabrication of WWFT; (**c**) XY side views of the fiber taper with overlap of 20 µm. The red circle represents the fiber taper location.

Next, two WWFTs are sequentially embedded into the sensing SMF to construct an intermodal interferometer, as shown in Figure 2. For the experimental test, the WWFT-based interferometer is connected with ASE light source (Golight Tech., ASE-C&L-10, Shenzhen, China) and CCD spectral module (Bayspec Inc., FBGA-1525-1610, Fremont, CA, America) for transmission spectrum acquisitions. The acquired spectra are then analyzed by using a LabVIEW program for Fast Fourier Transform (FFT), and the peak amplitudes/phases of the spectra after FFT are individually discussed. The schematic diagram of the experimental setup is shown in Figure 2.

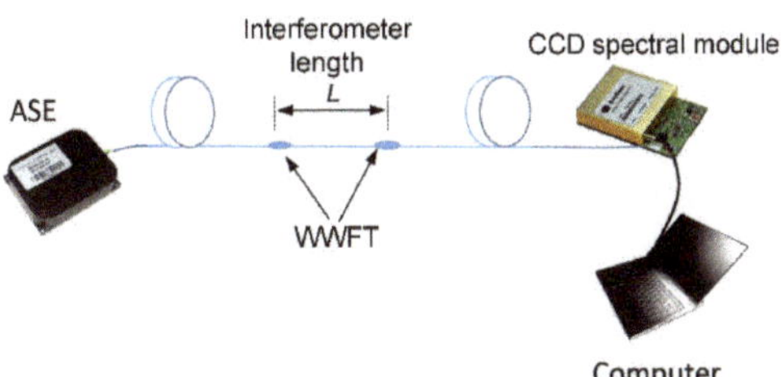

Figure 2. Experiment setup of WWFT-based interferometer. ASE: amplified spontaneous emission, CCD: charge-coupled device, WWFT: weak waist-enlarged fiber taper.

3. Spectral Characterization of WWFT-Based Interferometer

For the following analyses, sensors with different overlap values L_{vp} (10, 20 and 30 μm) are experimentally fabricated for comparison. Besides, various MZ interferometer samples with different interferometer lengths of L are also prepared. Figure 3 shows the transmission spectra of the WWFT-based interferometers with overlaps of 10 μm, 20 μm and 30 μm, respectively. We can observe that the induced insertion losses could be as low as 0.15 dB when L_{vp} = 20 μm. It increases slightly with the overlap distance enlarged since more high-order modes are excited and experience large transmission loss. From the figure, it also can be seen that if the overlap distance remains the same, the fringe contrast decreases as the interferometer length elongates. For example, when L_{vp} = 20 μm, the maximum fringe contrasts are 0.68 dB, 0.37 dB, 0.12 dB, respectively, for the interferometer lengths of 30.8 mm, 48.7 mm and 70.1 mm. Actually, when the interferometer length increases, the optical power ratio between sensing arm and reference arm of interferometer gradually decreases since the high-order mode suffers a higher transmission loss. Finally, it results in the deterioration of fringe contrast.

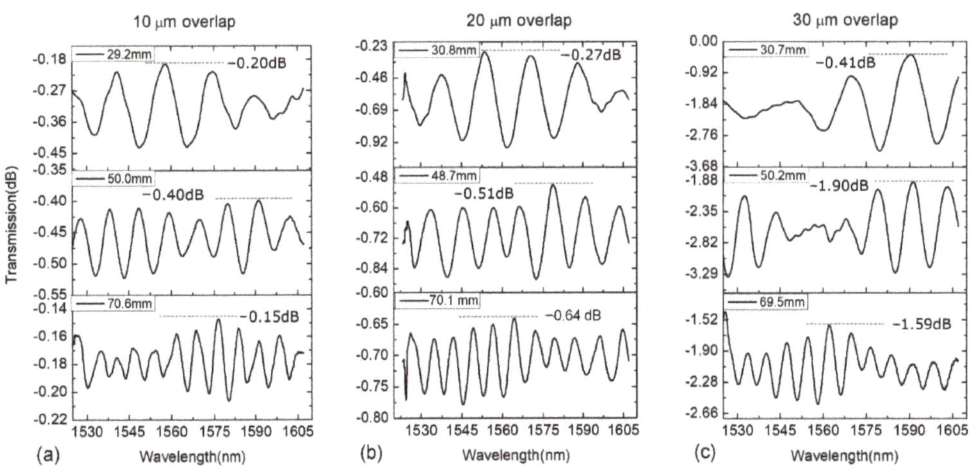

Figure 3. Transmission spectra of WWFT-based interferometers with different overlap values of (**a**) 10 μm, (**b**) 20 μm and (**c**) 30 μm, respectively.

Since WWFT is with non-adiabatic characteristics, part of the energy of LP_{01} is leaked and coupled into the high-order modes [21]. Next, in order to evaluate the specific mode order excited by WWFT, the Fast Fourier Transform (FFT) is applied to the transmission spectra in Figure 3. Herein, Figure 3b is selected, and another two sensors with interferometer lengths of 41.2 mm and 59.1 mm are supplemented for the following analyses. The results are shown in Figure 4a. For each interferometer with a certain physical length L, the optical path difference (OPD) δ could be calculated through the relationship of $\delta = cK_{Lm}/2\Delta v$ [22], where K_{Lm} is the index of peak amplitude location in the FFT spectra, and Δv is the whole frequency range of CCD spectral module. It should be noted that K_{lm} is a parameter with no dimension, and it varies with the frequency range of CCD spectral module. Hence, the calculated OPD itself is independent with the CCD module. Figure 4b illustrates the calculated OPD versus interferometer length L of the WWFT-based interferometer with different overlap distances of 10, 20 and 30 µm. It is well known that the optical path difference could be also expressed as $\delta = \Delta nL$, where Δn is the difference of effective refractive index between LP_{01} and the excited high-order mode. Therefore, associating with the two different expressions of δ above, the Δn could be determined in experiment, and the value is 2.59×10^{-3}, which is very close to the theoretical value of 2.73×10^{-3} between LP_{01} and LP_{11} mode using beam propagation method. Hence we can infer that the excited high-order mode is LP_{11} mode. The electric field distributions of LP_{01} and LP_{11} mode are displayed as the inset of Figure 4a.

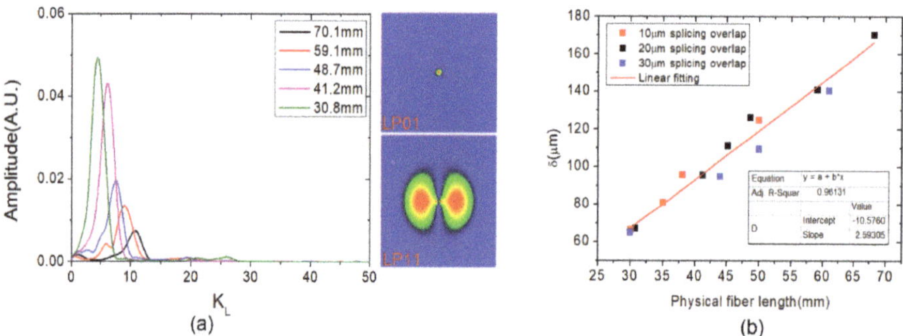

Figure 4. (**a**) Spectra after FFT for the WWFT-based interferometers with different interferometer lengths, inset: Electric field distributions of LP_{01} and LP_{11} mode; (**b**) Optical path difference versus interferometer length with overlaps of 10 µm, 20 µm and 30 µm, respectively.

4. Results and Analyses

Owing to the low insertion loss of WWFT, the proposed sensor can be applied for curvature multiplex sensing. Figure 5a shows the schematic of the curvature measurement for two cascaded WWFT-based interferometers. The fiber sensor 1# and 2# are with interference lengths L = 41.7 mm and 69.5 mm, respectively. Besides, the overlap distances are set as 20 µm and 30 µm, respectively, for the two sensors. For the two cascaded interferometers, the total insertion loss is measured as approximately 2.5 dB, and it can be further reduced with a smaller overlap value. Besides, the amount of multiplexed sensors depends on the measurable spectral range. According to the equation $\delta = cK_L/2\Delta v$ in the last paragraph, a larger spectral range results in a shorter optical path difference, and further, a higher resolution in the spectra after FFT, hence more sensors can be multiplexed. During the test, sensor 1# is placed between two supporting points with a separated distance of 115 mm, while sensor 2# keeps still. A one-dimensional translation stage with circular metal head is placed at the center of the sensor 1#. With the micrometer screw thread pushed forward, a fiber arc shape is formed. The curvature can be calculated by the expression of $\rho = 1/R = 8d/(4d^2 + D^2)$ [23], where R is the curvature radius of the fiber arc induced by pushing the one-dimension fiber translation stage, D is the distance of two supporting points, and d is the bending displacement. In our case, $d \ll D$, therefore, the curvature is simplified as $\rho = 1/R = 8d/D^2$.

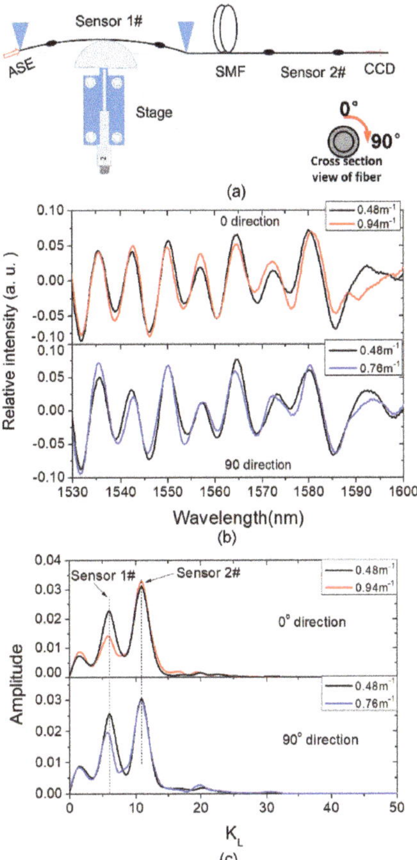

Figure 5. (**a**) Schematic of curvature test for the two cascaded WWFT-based interferometers; (**b**) Spectra and (**c**) Spectra after FFT for the WWFT-based interferometer 1# and 2# at 0° direction and 90° direction, respectively.

Based on the experimental setup in Figure 5a, different external curvatures are applied on the sensor 1#, and the corresponding transmission spectra at the two orthogonal directions (0° and 90°, schematic of fiber rotation angle is indicated in the inset of Figure 5a) are shown in Figure 5b. It should be noted that the axis Y in Figure 5b represents the relative intensity; it is obtained by normalizing the transmission spectrum amplitude with the input power, and then the gravity center of the spectral intensity shifts to zero to eliminate the DC component, hence both positive and negative values are shown. It can be seen that the wavelength shifts in Figure 5b are insignificant and irregular, which signifies that the curvature information cannot be directly demodulated. Figure 5c shows the corresponding spectra after FFT; it can be seen that the peak amplitudes of sensor 2# almost keep still, while the peak amplitudes of sensor 1# become lower as the curvature increases in both 0° and 90° directions. This phenomenon is reasonable since the LP_{11} mode is more sensitive to the bending effect, which leads to an enlarged transmission loss. Hence, the fringe contrast at the second WWFT point starts to deteriorate and further results in the reduction of peak amplitude.

Next, the continuous data acquisition is conducted, and the dynamic results are shown in Figure 6. The peak phases of spectra after FFT are continuously acquired as curvature varies between 0.48~0.94 m^{-1} (0° direction) °and 0.48~0.76 m^{-1} (90° direction). Figure 6a shows the phase shift when the WWFT-based fiber is placed at the initial position (defined as 0° direction). It can be seen that, when the curvature increases, the phase shift of sensor 1# is enlarged synchronously. Meanwhile, sensor 2# keeps almost unchanged, indicating that the curvature detections are independent between two sensors. Next, when the fiber sensor 1# is rotated to 90°, as shown in Figure 6b, we can see that the phase of sensor 1# shifts to the negative direction and presents different phase shifts when compared with those in 0° direction. This phenomenon signifies that the proposed sensor has diverse responses in two orthogonal directions, which can be further potentially applied for vector curvature sensing. Besides, for the 90° direction, we can also observe the cross-talk for sensor 2#. Theoretically speaking, no phase shift should occur for sensor 2# since no curvature is applied. This experimental phenomenon may be explained by the resolution ability of spectra after FFT, which is determined by the measurable spectral range of the CCD spectral module. As shown in Figure 5c, the peak amplitude of sensor 2# is easily affected by that of sensor 1# when the spectrum resolution ability is limited, and the cross-talk becomes more significant when the sensitivity decreases. Finally, the sensitivities in both 0° and 90° directions are shown in Figure 6c. The curvature sensitivities are determined to be 79.1°/m^{-1} and -48.0°/m^{-1}, respectively. The different curvature sensitivities are mainly due to the anisotropic distribution of the excited high-order mode (LP$_{11}$) in different axes. It should be explained that most of previous works relating to optical fiber curvature sensors are based on wavelength demodulation in unit of nanometer (nm) or amplitude demodulation in unit of decibel (dB). While in this study, the curvature sensing is realized based on phase demodulation in unit of degree (°). It should be noted that the wavelength and phase are two independent parameters which are inconvenient to convert to each other in our demodulation system. Hence, it is difficult to compare the curvature sensitivity performance presented in this study with the previously reported studies. Here, the sensitivity performance comparison is not discussed.

Finally, the temperature cross-sensitivity of the presented WWFT-based interferometer is discussed. The sensor is heated from 22.0 °C to 37.5 °C. Figure 7a shows the transmission spectra of the sensor with interferometer length of 41.2 mm and overlap value of 20 μm. We can hardly observe the wavelength shift with wavelength demodulation scheme. Then, based on phase demodulation scheme, the phase shift versus temperature is indicated in Figure 7b, and the temperature cross-sensitivity is determined to be 1.37 °/m^{-1}. When compared with the curvature sensitivities mentioned above, the temperature cross-sensitivity is insignificant. Besides, the phase fluctuation is monitored for 4 min at 22 °C, and the result is shown in Figure 7c. The recorded standard deviation is determined to be 0.071°. Here, we use three times the standard deviation as the phase resolution, i.e., 0.213°. Hence, the curvature resolutions of the proposed sensor are determined to be around $0.213/79.1 = 2.69 \times 10^{-3}$ m^{-1} at 0° direction and $0.213/48.0 = 4.44 \times 10^{-3}$ m^{-1} at 90° direction.

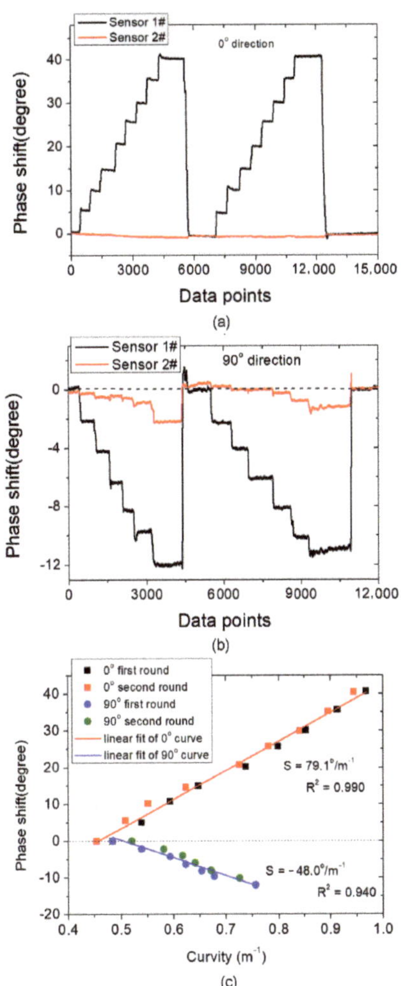

Figure 6. Phase responses of the WWFT-based interferometer 1# and sensor 2# in (**a**) 0° direction and (**b**) 90° direction; (**c**) Phase shift versus curvature in 0° and 90° direction.

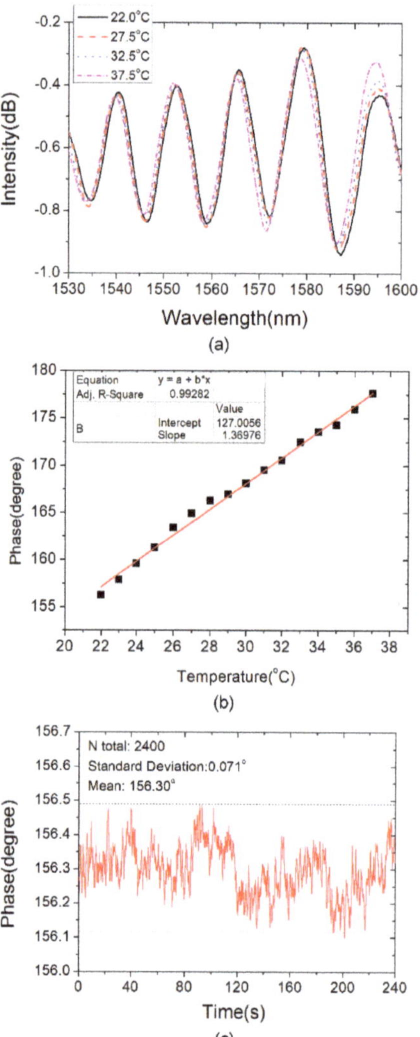

Figure 7. Temperature cross-sensitivity of WWFT-based interferometer: (**a**) Transmission spectra as temperature increases from 22 °C to 37.5 °C; (**b**) Phase shift versus temperature for the sensor with L = 41.2 mm and L_{vp} = 20 μm; (**c**) Continuous monitoring of phase fluctuation at 22.0 °C.

5. Conclusions

This study demonstrates a multiplexed WWFT-based curvature sensor and its rapid inline fabrication. When compared with the other types of fiber taper, the proposed WWFT has no difference in appearance with the SMF, which greatly reduces the insertion loss. The WWFTs could be rapidly embedded into the inline sensing fiber without the repeated cleaving-splicing process and splicing point, which enhances the solidity of sensor. Owing to the low transmission loss which could be as low as 0.15 dB, the WWFT-based interferometer can be further multiplexed based on phase-demodulation scheme. The experimental results verified that curvature variation can be dynamically detected for two cascaded WWFT-based interferometers. Besides, the proposed sensor shows diverse responses for the curvature changes in two orthogonal directions, and the sensitivities are determined to be $79.1°/m^{-1}$ and $-48.0°/m^{-1}$. Correspondingly, the curvature resolutions in two orthogonal directions are estimated as 2.69×10^{-3} m^{-1} and 4.44×10^{-3} m^{-1}, respectively. This result confirms the feasibility for vector curvature sensing applications in the future.

Author Contributions: Conceptualization, Y.G. and D.Y.; methodology, Y.G.; software, Y.G.; validation, L.W., X.L. and X.H.; formal analysis, D.Y.; investigation, L.W.; resources, Y.D.; data curation, D.Y.; writing—original draft preparation, D.Y. and Y.G.; writing—review and editing, D.Y and Y.G. All authors have read and agreed to the published version of the manuscript.

Funding: This work was supported, in part, by Stabilization Support Program for Higher Education Institutions of Shenzhen, grant number 20200811232156001, and, in part, by National Natural Science Foundation of China, grant number 51808347.

Institutional Review Board Statement: Not applicable.

Informed Consent Statement: Not applicable.

Conflicts of Interest: The authors declare no conflict of interest.

References

1. Zhu, M.; Xie, M.; Lu, X.; Okada, S.; Kawamura, S. A Soft Robotic Finger with Self-Powered Triboelectric Curvature Sensor Based on Multi-Material 3D Printing. *Nano Energy* **2020**, *73*, 104772. [CrossRef]
2. Xiong, C.; Lu, X.; Lin, X. Damage Assessment of Shear Wall Components for RC Frame–Shear Wall Buildings Using Story Curvature as Engineering Demand Parameter. *Eng. Struct.* **2019**, *189*, 77–88. [CrossRef]
3. Wang, L.; Xie, C.; Lin, Y.; Zhou, H.-Y.; Chen, K.; Cheng, D.; Dubost, F.; Collery, B.; Khanal, B.; Khanal, B.; et al. Evaluation and Comparison of Accurate Automated Spinal Curvature Estimation Algorithms with Spinal Anterior-Posterior X-Ray Images: The AASCE2019 Challenge. *Med Image Anal.* **2021**, *72*, 102115. [CrossRef] [PubMed]
4. Wang, S.; Wang, S.; Zhang, S.; Feng, M.; Wu, S.; Jin, R.; Zhang, L.; Lu, P. An Inline Fiber Curvature Sensor Based on Anti-Resonant Reflecting Guidance in Silica Tube. *Opt. Laser Technol.* **2019**, *111*, 407–410. [CrossRef]
5. Barrera, D.; Madrigal, J.; Sales, S. Long Period Gratings in Multicore Optical Fibers for Directional Curvature Sensor Implementation. *J. Lightwave Technol. JLT* **2018**, *36*, 1063–1068. [CrossRef]
6. Zhang, S.; Zhou, A.; Guo, H.; Zhao, Y.; Yuan, L. Highly Sensitive Vector Curvature Sensor Based on a Triple-Core Fiber Interferometer. *OSA Contin. OSAC* **2019**, *2*, 1953–1963. [CrossRef]
7. Marrujo-García, S.; Hernández-Romano, I.; Torres-Cisneros, M.; May-Arrioja, D.A.; Minkovich, V.P.; Monzón-Hernández, D. Temperature-Independent Curvature Sensor Based on In-Fiber Mach–Zehnder Interferometer Using Hollow-Core Fiber. *J. Lightwave Technol. JLT* **2020**, *38*, 4166–4173.
8. Zhang, C.; Zhao, J.; Miao, C.; Shen, Z.; Li, H.; Zhang, M. High-Sensitivity All Single-Mode Fiber Curvature Sensor Based on Bulge-Taper Structures Modal Interferometer. *Opt. Commun.* **2015**, *336*, 197–201. [CrossRef]
9. Gong, H.; Song, H.; Li, X.; Wang, J.; Dong, X. An Optical Fiber Curvature Sensor Based on Photonic Crystal Fiber Modal Interferometer. *Sens. Actuators A Phys.* **2013**, *195*, 139–141. [CrossRef]
10. Chen, E.; Dong, B.; Li, Y.; Wang, X.; Zhao, Y.; Xu, W.; Zhao, W.; Wang, Y. Cascaded Few-Mode Fiber down-Taper Modal Interferometers and Their Application in Curvature Sensing. *Opt. Commun.* **2020**, *475*, 126274. [CrossRef]
11. Xu, S.; Chen, H.; Feng, W. Fiber-Optic Curvature and Temperature Sensor Based on the Lateral-Offset Spliced SMF-FCF-SMF Interference Structure. *Opt. Laser Technol.* **2021**, *141*, 107174. [CrossRef]
12. Zhu, F.; Zhang, Y.; Qu, Y.; Su, H.; Jiang, W.; Guo, Y.; Qi, K. Fabry-Perot Vector Curvature Sensor Based on Cavity Length Demodulation. *Opt. Fiber Technol.* **2020**, *60*, 102382. [CrossRef]
13. Li, Y.; Chen, L.; Harris, E.; Bao, X. Double-Pass In-Line Fiber Taper Mach–Zehnder Interferometer Sensor. *IEEE Photonics Technol. Lett.* **2010**, *22*, 1750–1752. [CrossRef]
14. Zhang, P.; Liu, B.; Liu, J.; Xie, C.; Wan, S.; He, X.; Zhang, X.; Wu, Q. Investigation of a Side-Polished Fiber MZI and Its Sensing Performance. *IEEE Sens. J.* **2020**, *20*, 5909–5914. [CrossRef]
15. Shaklan, S.; Reynaud, F.; Froehly, C. Multimode Fiber-Optic Broad Spectral Band Interferometer. *Appl. Opt. AO* **1992**, *31*, 749–756. [CrossRef]
16. Geng, Y.; Li, X.; Tan, X.; Deng, Y.; Yu, Y. High-Sensitivity Mach–Zehnder Interferometric Temperature Fiber Sensor Based on a Waist-Enlarged Fusion Bitaper. *IEEE Sens. J.* **2011**, *11*, 2891–2894. [CrossRef]
17. Xu, F.; Li, C.; Ren, D.; Lu, L.; Lv, W.; Feng, F.; Yu, B. Temperature-Insensitive Mach-Zehnder Interferometric Strain Sensor Based on Concatenating Two Waist-Enlarged Fiber Tapers. *Chin. Opt. Lett. COL* **2012**, *10*, 070603.
18. Dash, J.N.; Jha, R.; Das, R. Enlarge-Tapered, Micro-Air Channeled Cavity for Refractive Index Sensing in SMF. *J. Lightwave Technol. JLT* **2019**, *37*, 5422–5427. [CrossRef]
19. Fu, H.; Zhao, N.; Shao, M.; Yan, X.; Li, H.; Liu, Q.; Gao, H.; Liu, Y.; Qiao, X. In-Fiber Quasi-Michelson Interferometer Based on Waist-Enlarged Fiber Taper for Refractive Index Sensing. *IEEE Sens. J.* **2015**, *15*, 6869–6874. [CrossRef]

20. An, J.; Jin, Y.; Sun, M.; Dong, X. Relative Humidity Sensor Based on SMS Fiber Structure with Two Waist-Enlarged Tapers. *IEEE Sens. J.* **2014**, *14*, 2683–2686. [CrossRef]
21. Snyder, A.W.; Love, J.D. *Optical Waveguide Theory*; Springer: Boston, MA, USA, 1984.
22. Ge, Y.; Wang, M.; Chen, X.; Rong, H. An optical MEMS pressure sensor based on a phase demodulation method. *Sens. Actuators A Phys.* **2008**, *143*, 224–229. [CrossRef]
23. Gong, Y.; Zhao, T.; Rao, Y.-J.; Wu, Y. All-Fiber Curvature Sensor Based on Multimode Interference. *IEEE Photonics Technol. Lett.* **2011**, *23*, 679–681. [CrossRef]

Article

An Ultra-Sensitive Multi-Functional Optical Micro/Nanofiber Based on Stretchable Encapsulation

Siheng Xiang, Hui You, Xinxiang Miao, Longfei Niu, Caizhen Yao , **Yilan Jiang and Guorui Zhou ***

Department of Engineering Optics, Research Center of Laser Fusion CAEP, Mianyang 621900, China; xiangsiheng1230@163.com (S.X.); yhworkyx@163.com (H.Y.); miaoxx@caep.cn (X.M.); niulf@caep.cn (L.N.); yaocaizhen2008@126.com (C.Y.); jiangyilan1023@163.com (Y.J.)
* Correspondence: zhougr@caep.cn; Tel.: +86-153-2811-4720

Abstract: Stretchable optical fiber sensors (SOFSs), which are promising and ultra-sensitive next-generation sensors, have achieved prominent success in applications including health monitoring, robotics, and biological–electronic interfaces. Here, we report an ultra-sensitive multi-functional optical micro/nanofiber embedded with a flexible polydimethylsiloxane (PDMS) membrane, which is compatible with wearable optical sensors. Based on the effect of a strong evanescent field, the as-fabricated SOFS is highly sensitive to strain, achieving high sensitivity with a peak gauge factor of 450. In addition, considering the large negative thermo-optic coefficient of PDMS, temperature measurements in the range of 30 to 60 °C were realized, resulting in a 0.02 dBm/°C response. In addition, wide-range detection of humidity was demonstrated by a peak sensitivity of 0.5 dB/% RH, with less than 10% variation at each humidity stage. The robust sensing performance, together with the flexibility, enables the real-time monitoring of pulse, body temperature, and respiration. This as-fabricated SOFS provides significant potential for the practical application of wearable healthcare sensors.

Keywords: optical micro/nanofiber; multi-functional; flexible encapsulation; wearable health-care sensor

Citation: Xiang, S.; You, H.; Miao, X.; Niu, L.; Yao, C.; Jiang, Y.; Zhou, G. An Ultra-Sensitive Multi-Functional Optical Micro/Nanofiber Based on Stretchable Encapsulation. *Sensors* **2021**, *21*, 7437. https://doi.org/10.3390/s21227437

Academic Editor: Jesús M. Corres

Received: 6 October 2021
Accepted: 3 November 2021
Published: 9 November 2021

Publisher's Note: MDPI stays neutral with regard to jurisdictional claims in published maps and institutional affiliations.

1. Introduction

At present, the research and application of flexible wearable electronics are reflected in various aspects of human life [1–4], such as electronic skin, wearable physiological monitoring and treatment devices, flexible conductive fabrics, thin-film transistors, and transparent thin-film flexible gate circuits [5–11]. The term 'flexible wearable electronics' generally refers to electronic devices or devices that have mechanical flexibility and can directly or indirectly adhere closely to the skin, including a group of functional devices that can respond to external stimuli via capacitive [12,13], resistive [14,15], piezoelectric [16,17], and triboelectric effects [18,19]. However, the need for higher sensitivity, faster response, and better anti-interference properties may push the limit of wearable electronic sensors due to the nature of low-frequency electromagnetic circuits. For example, drawbacks such as parasitic effects, and the risk of short circuit and electromagnetic disturbances, might restrict the practical applications of wearable electronics [20,21].

In addition to traditional electrical sensing, combining optical sensing with soft and stretchable packaging may provide an alternative means of developing cheap but reliable wearable sensors [22,23]. With respect to optical sensors, there are abundant attractive merits compared with conventional sensors, including ultra-high sensitivity, anti-electromagnetic interference, and excellent insulation, resulting in variation in the monitoring of environmental parameters [24]. For example, Nag et al. presented a new approach to the development of flexible transparent strain sensors which combined the use of a transparent fabric with a polydimethylsiloxane (PDMS) polymer using a layer-by-layer assembly process [25]. Yang et al. demonstrated, for the first time, a flexible optical fiber capable of sensing a wide range of body movements using the liquid PDMS-based tubular mold method [26]. Despite the success of these enlightening optical sensors, the intricate fabrication procedure of flexible and stretchable sensors makes commercial and practical usage difficult. As a combination of fiber optics and nanotechnology, optical micro/nanofibers (MNFs) have attracted increasing research interest due to their excellent properties, such as strong light field constraints, small bending radius, and high mechanical strength [27–29]. In particular, their strong evanescent fields ensure that MNFs are highly sensitive to tiny variations in the surroundings. It is worth noting that the as-fabricated MNFs, with their high performance and competitive scale, are vulnerable to the severe disturbances that occur in the application of wearable sensors. In order to employ MNFs in the development of reliable and robust stretchable optical fiber sensors (SOFSs), the packaging scheme must be well managed. For instance, Daniel et al. employed the change in the elastomer refractive index and modes confined within the fiber core when PDMS was pressed, fabricating a locally pressed etched optical fiber with a PDMS coating for sensing applications [30]. Additionally, Jesus et al. reported a compact and highly sensitive optical fiber temperature sensor based on the surface plasmon resonance effect, showing a linear response and a sensitivity of 2.6 nm/°C [31]. Despite the high performance of the above-mentioned sensors, a simple-to-fabricate, compact, hypersensitive, and multi-functional flexible optical fiber sensor remains a challenge for MNFs to some extent.

To address this issue, we propose a method to fabricate multi-functional stretchable fiber sensors by using a PDMS membrane embedded with MNFs. Based on the transition from guided modes to radiation modes of the waveguiding MNFs upon external stimuli, the as-fabricated MNF sensors are demonstrated to have outstanding sensitivity and repeatability when used in strain and temperature sensing. Moreover, by taking advantage of the water-absorbing ability of tin oxide, a wide detection range and highly sensitive humidity sensing are realized.

2. Materials and Methods

2.1. Fabrication and Manipulation of MNFs

In order to fabricate the MNFs, we used an electricity-heated mechanical stretching technique to draw a standard silica optical fiber into a biconical tapered fiber. In this method, a 3 cm-long section of the standard fiber was stripped of its cladding layer and fixed onto the MNF translation platform. When the fiber was heated to an optimal temperature, it was drawn into the horizontal plane until the diameter was reduced to the desired value. The entire MNF fabrication process was controlled by a computer program. To obtain uniformity among MNFs with different diameters, the temperature distribution in the drawing region and the speed of the stepper motors could be adjusted.

2.2. Fabrication of an SOFS

To fabricate an SOFS, we developed a sandwich-like structure to encapsulate the as-fabricated MNFs, which involved a three-step process: (1) 25 um of PDMS was prepared on a thin polyethylene (PET) flake to support the MNFs, which was firmly held in position using van der Waals forces and electrostatic interactions between the MNFs and the PDMS membrane; (2) 1 mL of degassed PDMS (mixing ratio of PDMS resin and curing agent was 10:1, Dow Corning Sylgard184) was slowly poured onto the PDMS/PET substrate, waiting for several seconds to ensure the substrate was covered with the fluid PDMS and that the MNFs were enclosed in the liquid; and (3) another thin PDMS/PET flake was placed on the substrate, followed by curing at 80 °C for 30 min, forming a 500 um-thick PDMS thin film. As for humidity sensing, a humidity-sensitive layer (tin oxide) was deposited using the magnetron sputtering method on the as-fabricated MNFs; the thickness of the tin oxide layer was about 20 nm for the sake of optimal sensing performance. According to the requirement of various sensing applications, the as-fabricated SOFS was peeled from the PET sheet.

2.3. Sensing Process of SOFS

Computer-controlled stepper motor translation platforms, which precisely control different step movements (including single step movement and periodic step movement), were prepared to measure the strain sensing performance of the SOFS. Before strain sensing, the as-fabricated SOFS was fastened onto the translation platforms with standard clamps and coupled with a 1550 nm wavelength fiber laser at one end. The optical output was detected by a spectrometer at the other end. The sensing process can be divided into two main parts: the first is a micro-strain measurement which contains several micron dimension step movements, and the other is an endurance test with hundreds of cycles of movements. Finally, in order to demonstrate the capability for health monitoring, the detection of cardiopalmus was conducted.

A constant temperature heating facility was used to test the temperature sensing performance of the SOFS, which could provide a stable temperature environment (ranging from 30 to 60 °C). During the temperature sensing process, the as-fabricated SOFS was positioned within the facility at all times and was always in a closed environment. Moreover, in order to demonstrate the capability of the SOFS for health monitoring, the detection of body temperature was conducted. A humidity generator and a detector were prepared to support the measurement of the SOFS's humidity sensing capability, which provided a continuous standard humidity atmosphere (ranging from 10 to 90% RH). Furthermore, for the duration of the humidity sensing process, the as-fabricated SOFS was placed in an airtight chamber. In order to demonstrate the capability for health monitoring, the detection of respiration was conducted.

2.4. Characterization Studies

An optical fiber laser (YMPSS-980-750-M-FBG, YM, Suzhou, China) and an optical fiber spectrometer (AQ6370D, YOKOGAWA, Japan) were used as a light source and a detector, respectively. A humidity generator (Su Zhou Hua Xiang Star Environmental Technology Co., Ltd., Suzhou, China) and a humidity detector (HC2A-DP, Rotronic) were used to generate standard humidity and measure the unknown environmental humidity, respectively. When a single-mode optical fiber (9/125, Corning, New York, NY, USA) was coupled with the fiber laser, the input optical power of the MNF was about 16 mW, with a stability of less than 0.05%. The stepper motors of the optical fiber drawing platform (V-508.932020, PI, Germany) and the translation stage (Zolix, China) were used to investigate the strain response of the sensor. A home-made system consisting of multiple channels was used for monitoring the response of the optical sensor to humidity, maintaining a relatively long recording period. A constant temperature heating facility (W-KW8-6-180, China) with a temperature resolution of 0.1 °C was provided to maintain a stable temperature in the range of 30 to 60 °C.

3. Results and Discussion

3.1. Concept and Principle of the SOFS

Although the conventional PDMS embedding method has been widely employed [32,33], demonstrating outstanding quality in the field of wearable sensors, there are still certain aspects of durability that should be considered in its application. Thus, in order to obtain a sufficiently robust SOFS, we added a pouring process to ensure that the SOFS was fastened firmly between the PDMS membranes. In this work, we employed a simple but dependable procedure to fabricate a sandwich-like SOFS—a schematic illustration is shown in Figure 1a, and detailed descriptions are displayed in the Materials and Methods section. Taking practical sensing objects into account, there may be little difference in the SOFS in terms of structure. Moreover, we chose PDMS to embed the MNFs because of its low refractive index (RI) compared with silica, flexibility, and biocompatibility. The stretchable PDMS film not only effectively fixes the MNFs and isolates the evanescent field but also maintains environmental stimuli transduced on MNFs with high sensitivity.

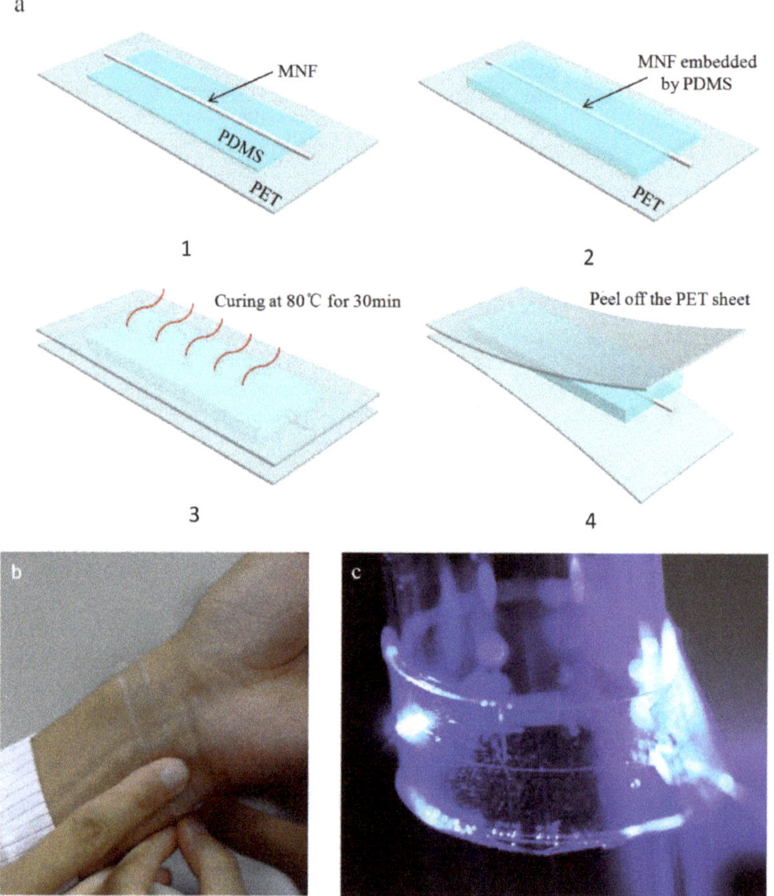

Figure 1. (**a**) Schematic illustration of the fabrication process of the sandwich-like SOFS. (**b**) Photograph of an SOFS attached to the wrist. (**c**) Photograph of an MNF-embedded PDMS patch pasted onto a glass tube.

Experimentally, highly uniform MNFs are fabricated through the tapered drawing of silica glass optical fibers, which allows for a guiding light at the micrometer scale. However, the as-fabricated MNFs, with a large proportion of evanescent fields exposed to the surroundings [34,35], are highly susceptible to environmental disturbance (e.g., pressure, vibration, and RI change) or contamination (e.g., dust and organic conglutination), which may lead to unpredictable perturbation of the guided signals. With the protection of the stretchable PDMS film, the above problems are able to be effectively suppressed to some extent, and an SOFS can be attached to the skin (Figure 1b) or pasted onto a glass cylinder with quite a large angle of bending (Figure 1c), exhibiting flexible and reliable characteristics in terms of wearable optical sensors.

Bending a flexible SOFS embedded with PDMS can lead to a bending-dependent transmission, which includes large-scale winding and tiny vibrations, enabling strain sensing. The RI of PDMS is a function of the temperature due to its large negative thermo-optic coefficient (about -1.0×10^{-4} per °C), and thus the guided modes of the MNFs are sensitive to the surrounding variation of RI owing to the large fractional evanescent fields outside the MNF, ensuring the SOFS's responsiveness to the change in temperature. Moreover, a wider expansion of aspects of sensing media could be compatible with the as-fabricated SOFS. Tin oxide, as a humidity-sensitive material, has already been used in humidity sensing [36–41]. Combining sensitive materials and MNFs embedded with PDMS could be a potential application in wearable sensors, such as in breathing masks and pulse monitors. Thus, the stretchable and flexible as-fabricated optical sensor offers promising capabilities for strain, temperature, and humidity sensing.

3.2. Strain Sensing

To investigate strain sensing, firstly, the SOFS embedded with a 1 um-diameter MNF was fastened to a computer-controlled stepper motor translation platform for bending sensing. By taking advantage of the high-precision moving stage, we designed each step to correspond to a replacement of 1 um, 2 um, and 5 um with 10 times bend and 10 times recover. As shown in Figure 2a–c, the MNF embedded with PDMS was manipulated using a micro-shift for strain response, which showed an approximately linear relationship between the output intensity and the applied strain (various degrees of bending). In order to exclude the disturbance of the light source and other factors, the displayed curves were detected using an optical fiber spectrometer at 1550 nm, which is the center of the wavelength of the light source, and where the lowest optical loss of such a single-mode optical fiber occurs. As for the 1 um per step test, with 10 times bend and recover movement, the response curve indicated a good reversibility of the sensor. Moreover, by connecting the points of the average data from every flat stage, the results present a highly linear relationship between the output intensity and the applied bending range. The relative intensity change $\Delta I/I_0$ (I_0 defined as the original output intensity of the SOFS) as a function of strain leads to a gauge factor (GF), where the gauge factor is defined as GF = $(\Delta I/I_0)$/strain. Meanwhile, the detectivity of the SOFS was shown to be highly sensitive, with a gauge factor of 118 and higher compared to conductive core–shell aramid nanofibrils (GF:18.8) [42], and silver nanowire elastomer nanocomposites (GF:14) [43]. Considering the fact that the limit of the spectrometer is 0.01 dBm, it is worth noting that nanometer-scale bend or micro-strain detection could be achieved. Compared with the above test, the 2 um and 5 um per step tests also manifested similar variation trends with a better response (GF:205 and GF:450, respectively), though with slightly weaker linearity, which is close to the recorded value (GF:1000) in cracking-assisted strain sensors [44]. To demonstrate the diameter optimization of the as-fabricated strain sensor, 3 um and 5 um-diameter SOFSs were used for the strain measurement, as shown in Figures S1 and S2. With regard to the 3 um-diameter SOFS, the response to strain (GF:64) was weaker than that of its 1 um counterpart, while the the 5 um-diameter SOFS was even more insensitive (GF:44), exhibiting no obvious response below a 20 um displacement. Thus, considering the practicability and sensitivity, the 1 um-diameter SOFS exhibited the best performance, showing potential for further research.

The long-term reversibility and stability of the SOFS were further investigated by automatically bending and recovering the sensor over 250 cycles using a computer-controlled moving platform. After the cycles, the sensor maintained a steady baseline intensity with a slight decrease of 5% (Figure 2d), which might be attributed to the hysteresis effect of PDMS or an error of the moving platform. In order to carry out a thorough inquiry of durability, more bending and recovering cycles were performed. The responses from approximately 1000 cycles are shown in Figure S3: with an increase in repeats, the response intensity of the SOFS decreases gradually. This attenuation of the intensity might be caused by the variation in the elastic property and/or a change in the refractive index of the PDMS membrane. To demonstrate the capability for health monitoring, the SOFS was attached to the wrist of a volunteer for the detection of cardiopalmus, which is a crucial aspect of physiological sensing. As shown in Figure 2e, compared with the contrast curve (maintaining the same bending level as the wrist), the response curve of the heartbeat exhibits a rapid fluctuation with a calculated 68 beats per minute.

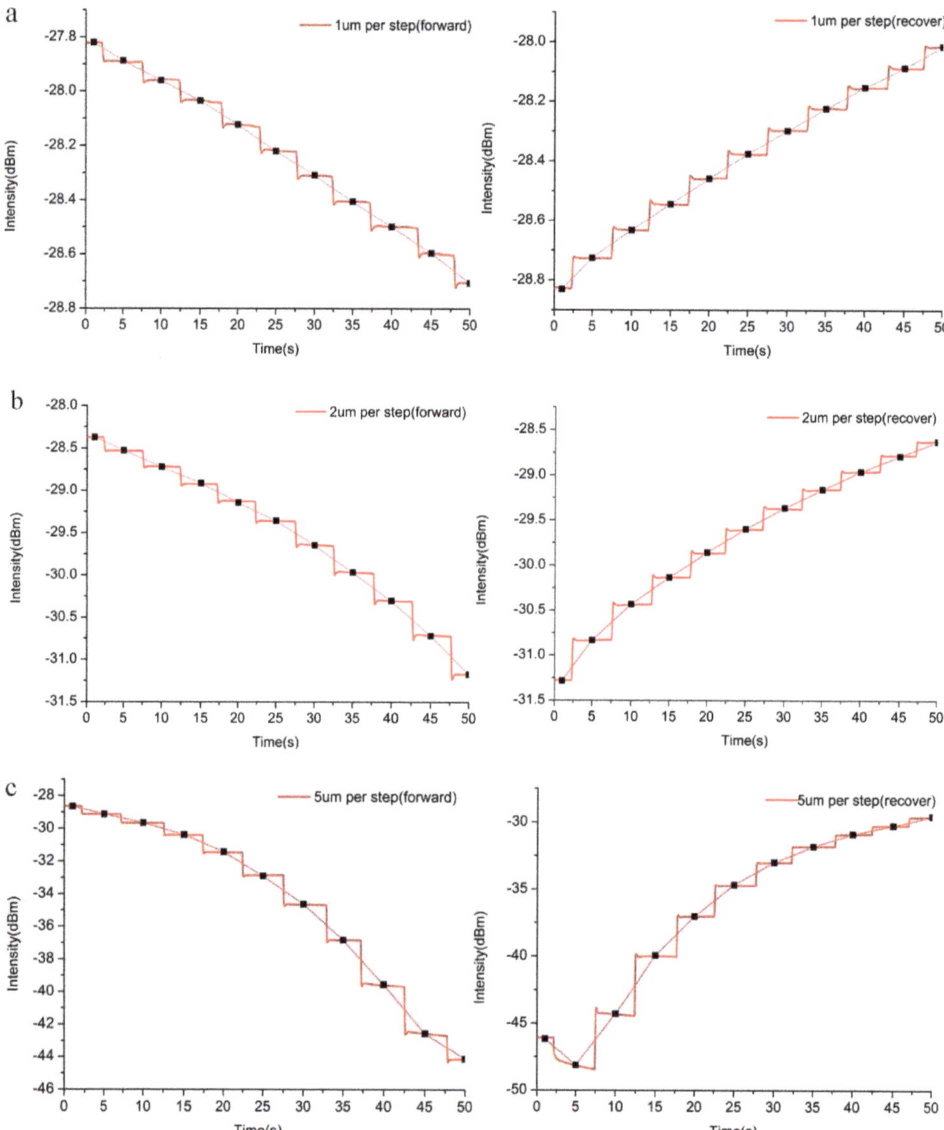

Figure 2. *Cont.*

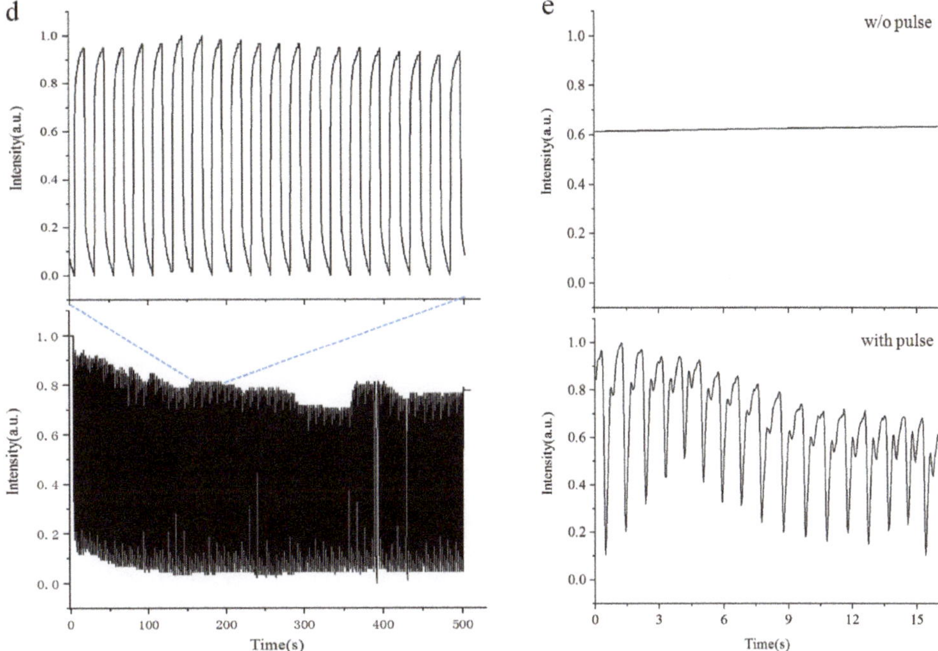

Figure 2. Characterization of the SOFS for strain sensing. Changes in optical intensity output derived from bend and recover tests. Each step corresponds to a displacement of (**a**) 1 μm, (**b**) 2 μm, and (**c**) 5 μm. (**d**) The durability test under a bend of 100 μm at a frequency of ~0.5 Hz. Inset: a magnifying view of the part of the response curve after 75 bend–recover cycles. (**e**) Measurement of the wrist pulse under normal conditions (about 68 beats per minute) and contrast conditions (maintaining the same bending extent without pulse).

3.3. Temperature Sensing

In addition to strain sensing, the refractive index characteristics of PDMS are promising for temperature sensing; it has a large negative thermo-optic coefficient, while the silica MNF shows a negligible change compared with that of PDMS. The SOFS with a 1 um-diameter MNF embedded in a 500 um-thick PDMS film was placed on a computer-controlled constant temperature heating facility, which provided a stable temperature in the range of 30 to 60 °C. The transmission spectra of the SOFS at different temperatures were recorded using a spectrometer (Figure 3a). The process of increasing the temperature was manipulated by a computer, which included 12 parts in all (part 1 increased the temperature from 30 to 35 °C, part 2 held the temperature at 35 °C for 90 s, part 3 increased the temperature from 35 to 40 °C, part 4 held the temperature at 40 °C for 90 s, and the rest were carried out in the same manner). With an increase in the temperature, the response curve indicated an approximately linear relationship between the output intensity and the applied temperature range. By defining the temperature sensitivity as $S = \Delta I / \Delta T$, where ΔI is the increase in the output intensity, and ΔT is the change in the temperature, the sensor achieved a sensitivity as high as 0.02 dBm per °C, which is similar to that of wearable temperature sensors based on metal oxides [45]. The decrease in transmission loss showed a tighter confinement of the guided light caused by the increased refractive index contrast between the PDMS and the MNF. In order to demonstrate its use as a wearable temperature sensor, the SOFS was directly pasted onto the back of the hand for temperature monitoring (the inset of Figure 3b). In order to avoid any disturbance caused by body motion, the subject tried to keep his hand as still as possible. To rule out changes in the ambient temperature, the whole process of measurement was conducted under a standard hundred-grade clean laboratory, maintaining a temperature of 25 °C and 50% RH all year. Figure 3b presents the typical response of the SOFS to a constant body temperature with remarkable stability, and no obvious change was observed during the period of detection.

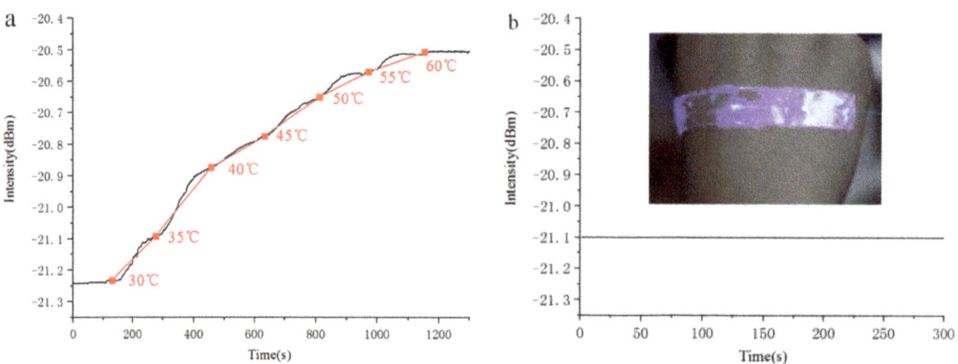

Figure 3. (**a**) Continuous temperature response of an SOFS measured in the range of 30–60 °C. (**b**) Real-time measurement of body temperature by attaching an SOFS to the back of the hand. Inset: a photograph of an SOFS attached to the back of the hand, which guided the 405 nm-wavelength laser for visualization.

3.4. Humidity Sensing

Besides the excellent stretchability and temperature response, an ultra-sensitive optical sensor with a wide range of humidity detection was obtained according to the Materials and Methods section. Interestingly, when we started to study the effect of different diameters on the humidity response, SOFSs with diameters that were too small or too large under the same thickness of the sensitive layer were demonstrated to be unresponsive. The phenomena of guided light scattering and optical absorption (conversion from light to heat, as shown in Figure S5) caused by the tin oxide layer are the main possible reasons for this; thus, balancing the MNF's parameters and the thickness of the sensitive layer appears to be of great importance to the humidity response. To investigate the humidity response, the SOFS embedded with an optimal 2 um-diameter MNF was placed in different humidity atmospheres, which was tested in an airtight cavity with a humidity generator and an electronic humidity detector.

Compared to the above sensors, the pattern of encapsulation was discriminating to some extent, which included a spindly silt (less than 0.5 cm^2) close to the most sensitive area in order to ensure wet air penetrated into the humidity-sensitive material. Tin oxide, a promising humidity-sensitive material, has not only been widely used in the field of humidity sensing but also popularized in the field of opto-electronics for decades [46–48]. It is worth noting that the morphology of tin oxide is characterized by its porous structure, which is beneficial for interacting with vapor and allowing vapor to transfer more smoothly. After combining with water vapor, the refractive index of tin oxide will change, which will affect the evanescent wave transmitted outside of the fiber as well. Figure 4a presents a continuous humidity measurement ranging from 10% to 90% RH using a home-made monitoring facility. At 10% to 30% RH, the as-fabricated sensor shows an inconspicuous response, probably due to the lack of enough water reacting with the tin oxide. Moreover, with the increase in humidity, the SOFS exhibits an excellent humidity sensing capability, especially at high concentrations of wet air (above 60% RH). Note that the robust sensitivity of the humidity sensor offers an opportunity for practical applications in daily life. By defining the humidity sensitivity as S = $\Delta I / \Delta H$, where ΔI is the increase in the output intensity, and ΔH is the change in the relative humidity, the as-fabricated sensor exhibits a peak sensitivity of 0.5 dB/% RH, which is far better than a humidity sensor based on a hetero-core optical fiber [49] and a no-core optical fiber [50], and an approach to humidity sensing based on a core-offset fiber Mach–Zehnder interferometer (S = 0.104 dB/% RH) [51]. Considering the durability of such a sensor, we also proceeded with more tests to demonstrate its repeatability, as shown in Figure 4b. A moderate fluctuation in the intensity response at every humidity stage was gauged with less than 10% variation. In particular, there is a better performance of the response within three times repeated measurements, which exhibits a flat curve (with less than 3% variation) of the output intensity except for the high-humidity conditions (90% RH). With the increase in repeated measurements, because continuous humidity testing from dry to wet might make the sensitive material saturated to a certain degree, the fluctuation in the intensity response becomes more or less larger under each of the humidity conditions.

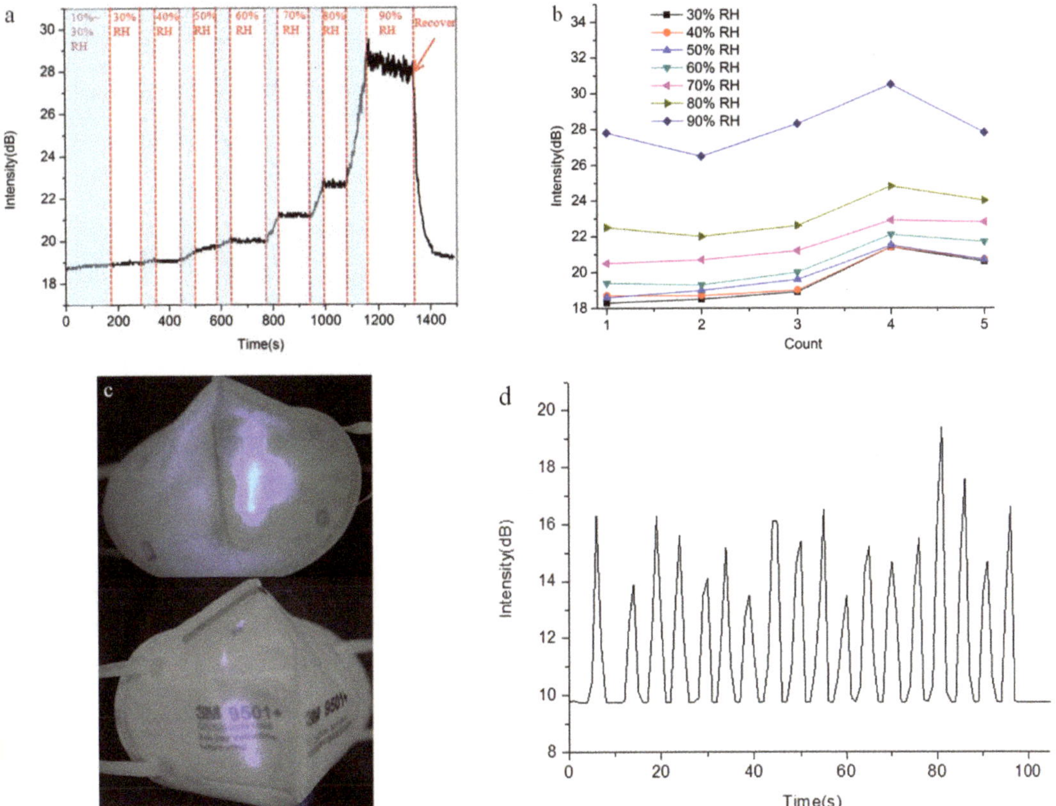

Figure 4. Characterization of the SOFS for humidity sensing. (**a**) Continuous humidity test ranging from 10% to 90% RH; the light blue area is the process of increasing the humidity. (**b**) Fluctuation in the intensity response at every humidity stage corresponding to five repeated measurements. (**c**) Photograph of a home-made breath-monitoring mask. (**d**) Real-time measurement of the respiration of a volunteer under deep breath (12 min^{-1}) conditions.

To exclude the disturbance of continuous humidity measurement and demonstrate the effect of the SOFS under all humidity conditions, the humidity stages from 30% to 90% RH were conducted individually. As shown in Figure S4, in this case, the holistic response tendency maintains that of its continuous counterpart. Moreover, under high-humidity conditions (greater than 60% RH), the output response displays an approximately symmetrical humidity and recovery cycle, while the recovery response at less than 60% RH shows a slightly weaker cycle, which might be attributed to the residual water penetrating into the porous tin oxide. To function as a wearable humidity sensor, the SOFS was directly integrated with a mask for the detection of respiration (Figure 4c), which is also a crucial indicator in the field of health monitoring. As shown in Figure 4d, under deep breath conditions within the breath-monitoring mask (12 min^{-1}), the humidity response curve presents quite a sharp peak and an excellent reverting capacity. In addition, the as-fabricated SOFS possesses a better response speed (of a few seconds) compared with its electronic counterpart (of minutes, especially for the recovery response) under a wide humidity range.

4. Conclusions

In summary, we have demonstrated an ultra-sensitive multi-functional optical micro/nanofiber embedded with a flexible encapsulation of PDMS, which possesses great potential in the field of wearable optical sensors. By taking advantage of the character of its strong evanescent field, the as-fabricated MNF embedded with a PDMS membrane is highly sensitive to weak bends, enabling a robust durability within hundreds of measurements and high sensitivity with a peak gauge factor of 450. At the same time, taking the large negative thermo-optic coefficient of PDMS into account, a temperature response in the range of 30 to 60 °C was realized, resulting in a resolution of 0.02 dBm/°C. In addition, a humidity measurement within a broad detection range from 30% to 90% RH was demonstrated, with a sensitivity of 0.5 dB/% RH, and less than 10% variation at each humidity stage. Moreover, the as-fabricated highly sensitive sensing performance with flexibility enables the real-time monitoring of pulse, body temperature, and respiration. The above results may pave the way towards future wearable optical sensors for health monitoring including real-time monitoring of pulse, respiration, and other aspects of life signs.

Supplementary Materials: The following are available online at https://www.mdpi.com/article/10.3390/s21227437/s1. Figure S1: Characterization of 3um parameter SOFS for strain sensing. Changes of output optical intensity derived from bend and recover a SOFS. Each step corresponding to a displacement of (a) 1 um, (b) 2 um, (c) 5 um, respectively. Figure S2: Characterization of 5um parameter SOFS for strain sensing. Changes of output optical intensity derived from bend and recover a SOFS. Each step corresponding to a displacement of (a) 1 um, (b) 2 um, respectively. Figure S3: Measurement of the durability test under a bend of 100 um at a frequency of 0.5 Hz for 1000 cycles in all. Figure S4: Humidity sensing of as-fabricated SOFS in different humidity and recover cycles, including (a) 30% RH and recover cycle, (b) 40% RH and recover cycle, (c) 50% RH and recover cycle, (d) 60% RH and recover cycle, (e) 70% RH and recover cycle, (f) 80% RH and recover cycle and (g) 90% RH and recover cycle, respectively. Figure S5: Infrared photograph of MNF deposited on tin oxide film, the red cross is the highest temperature area within MNF and the guided light direction is from left to right

Author Contributions: Conceptualization, S.X. and G.Z.; data curation, S.X. and H.Y.; formal analysis, S.X.; investigation, S.X., L.N. and C.Y.; methodology, S.X.; resources, Y.J. and G.Z.; software, H.Y.; supervision, X.M. and G.Z.; visualization, S.X.; writing—original draft, S.X.; writing—review and editing, S.X. All authors have read and agreed to the published version of the manuscript.

Funding: This project has received funding from the Youth Talents Foundation of the Research Center of Laser Fusion CAEP, China (RCFCZ5-2021-3), and The National Natural Science Foundation of China (12174355).

Institutional Review Board Statement: As the content of this study only involves the measurement of human signs, the measurement method and process will not cause any adverse reactions or potential risks to the subjects, so this study is not suitable for ethical review.

Informed Consent Statement: Informed consent was obtained from all subjects involved in the study.

Acknowledgments: We obtained approval to conduct human subject experiments from our department and from the volunteers before measuring wrist pulse, body temperature, and respiration rate.

Conflicts of Interest: There are no conflict to declare.

References

1. Wei, Z.; Lin, S.; Qiao, L.; Song, C.; Fei, W.; Tao, X.-M. Fiber-based wearable electronics: A review of materials, fabrication, devices, and applications. *Adv. Mater.* **2014**, *26*, 5310–5336.
2. Lv, T.; Yao, Y.; Li, N.; Chen, T. Wearable fiber-shaped energy conversion and storage devices based on aligned carbon nanotubes. *Nano Today* **2016**, *11*, 644–660. [CrossRef]
3. Mao, H.; Zhou, Z.; Wang, X.; Ban, C.; Huang, W. Polymer memory devices: Control of resistive switching voltage by nanoparticle-Decorated wrinkle interface. *Adv. Electron. Mater.* **2019**, *5*, 5–10.
4. Jost, K.; Dion, G.; Gogotsi, Y. Textile energy storage in perspective. *J. Mater. Chem. A* **2014**, *2*, 10776–10788. [CrossRef]
5. Bashir, T.; Ali, M.; Persson, N.K.; Ramamoorthy, S.K.; Skrifvars, M. Stretch sensing properties of knitted structures made of PEDOT-coated conductive viscose and polyester yarns. *Text. Res. J.* **2014**, *84*, 323–333. [CrossRef]
6. Nilsson, E.; Rigdahl, M.; Hagström, B. Electrically conductive polymeric bi-component fibers containing a high load of low-structured carbon black. *J. Appl. Polym. Sci.* **2015**, *132*, 42255–42264. [CrossRef]
7. Hong, C.H.; Ki, S.J.; Jeon, J.H.; Che, H.L.; Park, I.K.; Kee, C.D.; Oh, I.K. Electroactive bio-composite actuators based on cellulose acetate nanofibers with specially chopped polyaniline nanoparticles through electrospinning. *Compos. Sci. Technol.* **2013**, *87*, 135–141. [CrossRef]
8. Liu, X.; Guo, R.; Shi, Y.; Deng, L.; Li, Y. Durable, washable, and flexible conductive PET fabrics designed by fiber interfacial molecular engineering. *Macromol. Mater. Eng.* **2016**, *301*, 1383–1389. [CrossRef]
9. Karim, N.; Zhang, M.; Afroj, S.; Koncherry, V.; Potluri, P.; Novoselov, K.S. Graphene-based surface heater for de-icing applications. *RSC Adv.* **2018**, *8*, 16815–16823. [CrossRef]
10. Tadesse, M.G.; Mengistie, D.A.; Chen, Y.; Wang, L.; Loghin, C.; Nierstrasz, V. Electrically conductive highly elastic polyamide/lycra fabric treated with PEDOT:PSS and polyurethane. *J. Mater. Sci.* **2019**, *54*, 9591–9603. [CrossRef]
11. Szymon, M.; Wardak, C.; Pietrzak, K. Effect of multi-walled carbon nanotubes on analytical parameters of laccase-based biosensors received by soft plasma polymerization technique. *IEEE Sens. J.* **2020**, *20*, 8423–8428. [CrossRef]
12. Mannsfeld, S.C.; Tee, B.C.; Stoltenberg, R.M.; Chen, C.M.; Barman, S.; Muir, B.V.; Sokolov, A.N.; Reese, C.; Bao, Z. Highly sensitive flexible pressure sensors with microstructured rubber dielectric layers. *Nat. Mater.* **2010**, *9*, 859–864. [CrossRef]
13. Xu, H.; Lv, Y.; Qiu, D.; Zhou, Y.; Zeng, H.; Chu, Y. An ultra-stretchable, highly sensitive and biocompatible capacitive strain sensor from an ionic nanocomposite for on-skin monitoring. *Nanoscale* **2019**, *11*, 1570–1578. [CrossRef] [PubMed]
14. Jason, N.N.; Ho, M.D.; Cheng, W. Resistive electronic skin. *J. Mater. Chem. C* **2017**, *5*, 5845–5866. [CrossRef]
15. Kim, K.H.; Jang, N.S.; Ha, S.H.; Cho, J.H.; Kim, J.M. Fabrication of metal nanowire based stretchable mesh electrode for wearable heater application. *Small* **2018**, *14*, 575–581.
16. Zhu, M.; Shi, Q.; He, T.; Yi, Z.; Ma, Y.; Yang, B.; Chen, T.; Lee, C. Self-powered and self-functional cotton sock using piezoelectric and triboelectric hybrid mechanism for healthcare and sports monitoring. *ACS Nano* **2019**, *13*, 1940–1952. [CrossRef]
17. Zhao, G.; Zhang, X.; Cui, X.; Wang, S.; Liu, Z.; Deng, L.; Qi, A.; Qiao, X.; Li, L.; Pan, C.; et al. Piezoelectric polyacrylonitrile nanofiber film-based dual-function self-powered flexible sensor. *ACS Appl. Mater. Interfaces* **2018**, *10*, 15855–15863. [CrossRef] [PubMed]
18. Ha, M.; Lim, S.; Cho, S.; Lee, Y.; Na, S.; Baig, C.; Ko, H. Skin-inspired hierarchical polymer architectures with gradient stiffness for spacer-free, ultrathin, and highly sensitive triboelectric sensors. *ACS Nano* **2018**, *12*, 3964–3974. [CrossRef]
19. Wu, F.; Li, C.; Yin, Y.; Cao, R.; Li, H.; Zhang, X.; Zhao, S.; Wang, J.; Wang, B.; Xing, Y.; et al. Fabrication of large-area bimodal sensors by all-inkjet-printing. *Adv. Mater. Technol.* **2019**, *4*, 1800703–1800712.
20. Miller, D.A.B. Rationale and challenges for optical interconnects to electronic chips. *Proc. IEEE* **2000**, *88*, 728–749. [CrossRef]
21. Wang, C.; Hwang, D.; Yu, Z.; Takei, K.; Park, J.; Chen, T.; Ma, B.; Javey, A. User-interactive electronic skin for instantaneous pressure visualization. *Nat. Mater.* **2013**, *12*, 899–904. [CrossRef]
22. Guo, J.; Liu, X.; Jiang, N.; Yetisen, A.K.; Yuk, H.; Yang, C.; Khademhosseini, A.; Zhao, X.; Yun, S.-H. Highly stretchable, strain sensing hydrogel optical fibers. *Adv. Mater.* **2016**, *28*, 10244–10249. [CrossRef] [PubMed]
23. Leal-Junior, A.; Frizera, A.; Lee, H.; Mizuno, Y.; Nakamura, K.; Paixão, T.; Leitão, C.; Domingues, M.F.; Alberto, N.; Antunes, P.; et al. Strain, temperature, moisture, and transverse force sensing using fused polymer optical fibers. *Opt. Express* **2018**, *26*, 12939–12947. [CrossRef]
24. Rantala, J.; Haennikaeinen, J.; Vanhala, J. Fiber optic sensors for wearable applications. *Pers. Ubiquitous Comput.* **2011**, *15*, 85–96. [CrossRef]
25. Nag, A.; Simorangkir, R.; Valentin, E.; Bjorninen, T.; Ukkonen, L.; Hashmi, R.M.; Mukhopadhyay, S.C. A transparent strain sensor based on PDMS-embedded conductive fabric for wearable sensing applications. *IEEE Access* **2018**, *4*, 71020–71027. [CrossRef]
26. Optical Society of America. *In a First for Wearable Optics, Researchers Develop Stretchy Fiber to Capture Body Motion*; Optical Society of America: Washington, DC, USA, 2017.
27. Tong, L. Micro/Nanofibre optical sensors: Challenges and prospects. *Sensors* **2018**, *18*, 903. [CrossRef] [PubMed]
28. Wu, Y.; Yao, B.; Yu, C.; Rao, Y. Optical graphene gas sensors based on microfibers: A review. *Sensors* **2018**, *18*, 941. [CrossRef]
29. Gong, Y.; Yu, C.-B.; Wang, T.-T.; Liu, X.-P.; Wu, Y.; Rao, Y.-J.; Zhang, M.-L.; Wu, H.-J.; Chen, X.-X.; Peng, G.-D. Highly sensitive force sensor based on optical microfiber asymmetrical fabry-perot interferometer. *Opt. Express* **2014**, *22*, 3578–3584. [CrossRef]
30. Kacik, D.; Tatar, P.; Turek, I. Locally pressed etched optical fiber with PDMS coating for a sensor application. *Optik* **2016**, *127*, 5631–5635. [CrossRef]

31. Velazquez-Gonzalez, J.S.; Monzon-Hernandez, D.; Martinez-Pinon, F.; May-Arrioja, D.A.; Hernanderz-Romano, I. Surface plasmon resonance-based optical fiber embedded in PDMS for temperature sensing. *IEEE J. Sel. Top. Quantum Electron.* **2017**, *12*, 126–131. [CrossRef]

32. Dai, X.; Schriemer, H.P.; Kleiman, R.N.; Ding, H.; Blanchetiere, C.; Jacob, S.; Mihailov, S.J. Chemical sensor using polymer coated tapered optical fibers. *Int. Soc. Opt. Photonics* **2010**, *7750*, 77500–77506.

33. Wang, D.; Sheng, B.; Peng, L.; Huang, Y.; Ni, Z. Flexible and optical fiber sensors composited by graphene and pdms for motion detection. *Polymers* **2019**, *11*, 1433. [CrossRef]

34. Zhao, C.; Hou, L.; Xu, B.; Chen, H.; Wang, D. High-sensitivity hydraulic pressure sensor realized with PDMS film-based Fabry-Perot interferometer. In Proceedings of the 17th International Conference on Optical Communications and Networks, Zhuhai, China, 16–19 November 2018.

35. Tong, L.; Lou, J.; Mazur, E. Single-mode guiding properties of subwavelength-diameter silica and silicon wire waveguides. *Opt. Express* **2004**, *12*, 1025–1035. [CrossRef]

36. Ismail, A.S.; Mamat, M.H.; Malek, M.F.; Yusoff, M.M.; Mohamed, R.; Md. Sin, N.D.; Suriani, A.B.; Rusop, M. Heterogeneous SnO_2/ZnO nanoparticulate film: Facile synthesis and humidity sensing capability. *Mater. Sci. Semicond. Process.* **2018**, *81*, 127–138. [CrossRef]

37. Ismail, A.S.; Mamat, M.H.; Yusoff, M.M.; Malek, M.F.; Zoolfakar, A.S.; Rani, R.A.; Suriani, A.B.; Mohamed, A.; Ahmad, M.K.; Rusop, M. Enhanced humidity sensing performance using Sn-Doped ZnO nanorod Array/SnO_2 nanowire heteronetwork fabricated via two-step solution immersion. *Mater. Lett.* **2018**, *210*, 258–262. [CrossRef]

38. Lin, C.; He, J.; Jiang, S.; Liu, X.; Zhao, D.; Zhang, D. Humidity sensing characteristics of tin oxide thin film gas sensors varying with the operating voltage. In Proceedings of the IEEE 11th International Conference on Solid-State and Integrated Circuit Technology, Xian, China, 29 October–1 November 2012.

39. Krishnakumar, T.; Jayaprakash, R.; Singh, V.N.; Mehta, R.B.; Phani, A.R. Synthesis and characterization of tin oxide nanoparticle for humidity sensor applications. *J. Nano Res.* **2008**, *4*, 91–101. [CrossRef]

40. Duraia, E.; Das, S.; Beall, G.W. Humic acid nanosheets decorated by tin oxide nanoparticles and there humidity sensing behavior. *Sens. Actuators* **2019**, *280*, 210–218. [CrossRef]

41. Ascorbe, J.; Corres, J.M.; Matias, I.R.; Arregui, F.J. High sensitivity humidity sensor based on cladding-etched optical fiber and lossy mode resonances. *Sens. Actuators B Chem.* **2016**, *233*, 7–16. [CrossRef]

42. Xiang, S.; Han, L.; Lv, D.; Yu, X.; Wu, C. Conductive core-shell aramid nanofibrils: Compromising conductivity with mechanical robustness for organic wearable sensing. *ACS Appl. Mater. Interfaces* **2018**, *11*, 3466–3473.

43. Rahimi, R.; Ochoa, M.; Yu, W.; Ziaie, B. Highly stretchable and sensitive unidirectional strain sensor via laser carbonization. *ACS Appl. Mater. Interfaces* **2015**, *7*, 4463–4470. [CrossRef] [PubMed]

44. Chen, S.; Wei, Y.; Wei, S.; Lin, Y.; Liu, L. Ultrasensitive cracking-assisted strain sensors based on silver nanowires/graphene hybrid particles. *ACS Appl. Mater. Interfaces* **2016**, *8*, 25563–25570. [CrossRef]

45. Shin, J.; Jeong, B.; Kim, J.; Nam, V.B.; Yoon, Y.; Jung, J.; Hong, S.; Lee, H.; Eom, H.; Yeo, J. Sensitive wearable temperature sensor with seamless monolithic integration. *Adv. Mater.* **2020**, *32*, 1905527. [CrossRef] [PubMed]

46. Rasi, D.D.C.; van Thiel, P.M.J.G.; Bin, H.; Hendriksl, K.H.; Heintges, G.H.L.; Wienk, M.M.; Becker, T.; Li, Y.; Riedl, T.; Janssen, R.A.J. Solution-processed tin oxide-pedot:pss interconnecting layers for efficient inverted and conventional tandem polymer solar cells. *Sol. RRL* **2019**, *3*, 366–376.

47. Wang, X.; Zhao, Z.; Jun, X.; Xiong, J.; Chen, Q. Perovskite solar cells based on a spray-coating tin oxide film. *Chin. Opt.* **2019**, *12*, 1040–1047. [CrossRef]

48. Way, A.; Luke, J.; Evans, A.D.; Li, Z.; Tsoi, W.C. Fluorine doped tin oxide as an alternative of indium tin oxide for bottom electrode of semi-transparent organic photovoltaic devices. *AIP Adv.* **2019**, *9*, 085220–085229. [CrossRef]

49. Akita, S.; Sasaki, H.; Watanabe, K.; Seki, A. A humidity sensor based on a hetero-core optical fiber. *Sens. Actuators B Chem.* **2010**, *147*, 385–391. [CrossRef]

50. Xia, L.; Li, L.; Li, W.; Kou, T.; Liu, D. Novel optical fiber humidity sensor based on a no-core fiber structure. *Sens. Actuator A Phys.* **2013**, *190*, 1–5. [CrossRef]

51. Shuai, L.; Meng, H.; Rui, X.; Deng, S.; Wang, X.; Jiao, T.; Tan, C.; Huang, X. Humidity sensor based on core-offset fiber mach-zehnder interferometer coated graphene oxide film. In Proceedings of the 2017 16th International Conference on Optical Communications and Networks (ICOCN), Wuzhen, China, 7–10 August 2017.

sensors

MDPI

Article

Broadband Acoustic Sensing with Optical Nanofiber Couplers Working at the Dispersion Turning Point

Xu Gao [1,2], Jiajie Wen [2], Jiajia Wang [3] and Kaiwei Li [4,*]

[1] College of Instrumentation & Electrical Engineering, Jilin University, Changchun 130000, China; gaox19870513@163.com
[2] School of Opto-Electronic Engineering, Changchun University of Science and Technology, Changchun 130022, China; wenjiajie9504@163.com
[3] College of Agricultural Equipment Engineering, Henan University of Science and Technology, Luoyang 471003, China; jjw@haust.edu.cn
[4] Key Laboratory of Bionic Engineering of Ministry of Education, Jilin University, Changchun 130022, China
* Correspondence: kaiwei_li@jlu.edu.cn

Abstract: Herein, a broadband ultrasensitive acoustic sensor based on an optical nanofiber coupler (ONC) attached to a diaphragm is designed and experimentally demonstrated. The ONC is sensitive to axial strain and works as the core transducing element to monitor the deformation of the diaphragm driven by acoustic waves. We first theoretically studied the sensing property of the ONC to axial strain and the deformation of the diaphragm. The results reveal that ONC working at the dispersion turning point (DTP) shows improved ultra-sensitivity towards axial strain, and the largest deformation of the circular diaphragm occurs at the center. Guided by the theoretical results, we fabricated an ONC with a DPT at 1550 nm, and we fixed one end of the ONC to the center of the diaphragm and the other end to the edge to construct the acoustic sensor. Finally, the experimental results show that the sensor can achieve accurate measurement in the broadband acoustic wave range of 30~20,000 Hz with good linearity. Specifically, when the input acoustic wave frequency is 120 Hz, the sensitivity reaches 1923 mV/Pa, the signal-to-noise ratio is 42.45 dB, and the minimum detectable sound pressure is 330 μPa/Hz$^{1/2}$. The sensor has the merits of simple structure, low cost, and high performance, and it provides a new method for acoustic wave detection.

Keywords: optical fiber coupler; acoustic sensing; dispersion turning point

Citation: Gao, X.; Wen, J.; Wang, J.; Li, K. Broadband Acoustic Sensing with Optical Nanofiber Couplers Working at the Dispersion Turning Point. *Sensors* **2022**, *22*, 4940. https://doi.org/10.3390/s22134940

Academic Editor: Karim Benzarti

Received: 8 June 2022
Accepted: 28 June 2022
Published: 30 June 2022

Publisher's Note: MDPI stays neutral with regard to jurisdictional claims in published maps and institutional affiliations.

1. Introduction

Acoustic wave sensors play a dominant role in medical diagnosis, earthquake prediction, and volcano monitoring. Compared with the traditional magnetoelectric acoustic sensors, optical fiber acoustic sensors have the advantages of small size, light weight, high sensitivity, and strong anti-electromagnetic interference ability [1]. In recent years, researchers have proposed many acoustic sensing technologies based on the principle of optical fiber sensing. Depending on differences in the sensor configuration and mechanism of acoustic-to-optical signal conversion, these sensors can be divided into two categories. One is the fiber-optic Fabry–Pérot cavity, which is constructed of a fiber tip and a reflective diaphragm that can convert acoustic waves to changes in optical cavity length and, hence, the shift in the optical interference signal via its mechanical deformation [2–7]. The other type is based on a strain sensor that is attached to a diaphragm/plate and is responsible for converting the acoustic vibration to the axial elongation of the fiber optic strain sensor and, hence, the optical signal [8–11].

In the past decades, a variety of diaphragm materials, such as silicon [2], polymer [3], graphene [4,5], chitosan [6], and silver [7], have been studied for fiber-optic Fabry–Pérot cavity acoustic sensors. In addition, reducing the thickness-to-diameter ratio of the diaphragm can significantly improve the sensitivity of such acoustic sensors [1]. However, the production processes of polymer film, metal film, and micromachined silicon film are complex and expensive, involving mechanical spinning, chemical etching, or micromachining. Further, the requirements for demodulation are relatively high and costly.

The second type of fiber optic acoustic sensor has also undergone rapid development in recent years. Sumit Dass et al. designed an acoustic sensor based on a single-mode tapered fiber and a butyronitrile thin diaphragm, achieving a minimum detectable sound pressure of 21.11 Pa/Hz$^{1/2}$ and flat output response at frequencies of 250 Hz~2500 Hz [8]. Shun Wang et al. proposed an optical fiber acoustic sensor based on a nonstandard fused coupler and aluminum foil, which achieved a sensitivity of up to 2.63 mW/Pa and an acoustic measurement range of 20 Hz~20 kHz [9]. Wenjun Ni et al. proposed a thin-core ultra-long-period fiber-grating-based acoustic wave sensor showing an operation range of 1 Hz–3 kHz and high sensitivity up to 1890 mv/Pa at 5 Hz [10]. However, some of the above-mentioned acoustic wave sensors cannot achieve wide-band acoustic wave measurement. Also, the sensitivities are usually relatively low, and preamplifiers are required to achieve high sensitivity. Therefore, exploiting new optical fiber acoustic sensors with broad working frequency ranges and high sensitivity still attracts the wide attention of researchers. Recently, Kaijun Liu et al. demonstrated that the microfiber Mach–Zehnder interferometer working at the dispersion turning point (DTP) can be employed for vibration detection, and ultra-high sensitivity was achieved [11].

We recently discovered and demonstrated the dispersion turning point (DTP) in optical nanofiber couplers (ONCs), which can improve the sensing performance [12,13]. ONCs working at the DTP show enhancements of nearly two orders of magnitude compared with conventional fiber coupler sensors for refractive index (RI) sensing [12,14], temperature sensing [13], biochemical sensing [15], and axial strain sensing [16]. This study further implements the ONC with a DTP in acoustic sensing. The acoustic sensor was constructed by fixing the fiber coupler on a specially designed sensing diaphragm. Acoustic waves drive the sensing diaphragm to vibrate, and the vibration changes the coupling length of the ONC dynamically and is measured by the ONC, which works as an ultra-sensitive tensile sensor. We first theoretically calculated and analyzed the strain sensing performance of the ONC operation at the DTP. Then, we numerically studied the acoustic induced vibration characteristics of the polyethylene (PE) diaphragm. Finally, we experimentally realized a sensing device that can achieve acoustic wave measurement of 30–20 kHz and achieved a high sensitivity of 1929 mV/Pa at 120 Hz. The sensor has a simple structure, low requirements for demodulation equipment, and low cost, which provides a new method for acoustic wave detection.

2. Principle of Acoustic Sensing Using the ONC–Diaphragm Configuration

The configuration of the proposed ONC-based acoustic sensor is shown in Figure 1a. An ONC that operates at the DTP was adopted as the core sensing element, with one end fixed at the center of a PE diaphragm and the other end fixed on the edge of the diaphragm. As shown in Figure 1b, the dynamic acoustic wave drives the diaphragm to vibrate accordingly, and the vibration imposes stress on the ONC dynamically, leading to variation in the coupling length L. As the ONC sensor working at the DTP is quite sensitive to changes in length, the acoustic wave of slight sounds can be readily detected by tracing the optical output signal of the ONC sensor.

2.1. Working Principle of the ONC

As the sensing of acoustic waves is realized by stretching the fiber coupler, we first analyze the strain sensing performance of the ONC numerically. The configuration of the ONC is shown in Figure 2. The ONC is formed by fusion splicing two single-mode fibers, including a coupling section, two input ports, and two output ports. The coupling section of the two microfibers forms a new waveguide. When light enters the conical transition region from Port 1, both the even mode and odd mode can be excited simultaneously. The two modes propagate along the waveguide and gradually accumulate phase difference, finally coupling to the two output ends to produce mode interference. When the coupling section is stretched, the length of the optical path varies and the interference fringes of the output spectrum shift.

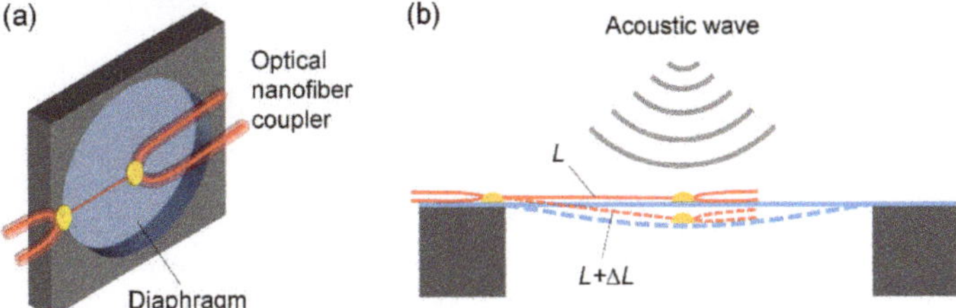

Figure 1. (**a**) Structure diagram of the optical-nanofiber-based acoustic sensor. (**b**) The acoustic wave sensing principle of the diaphragm-supported optical nanofiber sensor.

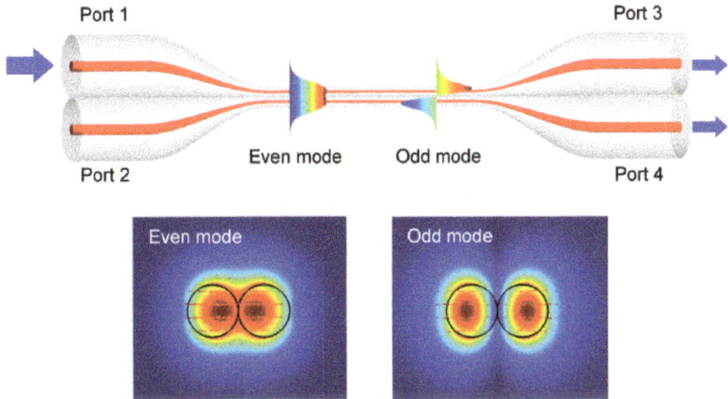

Figure 2. Schematic diagram of the nanofiber coupler and the modal field distributions for the even mode and odd mode.

Assume that the input optical power at Port 1 is P_0 at a certain wavelength λ_N. Thus, the output energy of output Ports 3 and 4 can be obtained as [13]:

$$P_3 = P_0 \cos^2\left(\frac{1}{2}\phi\right), \tag{1}$$

$$P_4 = P_0 \sin^2\left(\frac{1}{2}\phi\right) \tag{2}$$

The phase difference ϕ between the two modes satisfies:

$$\phi_N = \frac{2\pi L \left(n_{\text{eff}}^{\text{even}} - n_{\text{eff}}^{\text{odd}}\right)}{\lambda_N} = (2N - 1)\pi \tag{3}$$

where $n_{\text{eff}}^{\text{even}}$ and $n_{\text{eff}}^{\text{odd}}$ are the effective refractive index values (ERIs) of the odd mode and even mode. L is the length of the uniform coupling section, and N is an integer. In this study, we only consider the uniform waist region as it is much thinner and much longer than the tapered region, and the waist region is much more sensitive than the tapered region.

When external stretching is applied to the nanofiber coupler, the waist segment undergoes a prolongation along the axial direction and shrinkage in the radial direction. Concurrently, the RI of the waist region also undergoes a decrease as a result of the elasto-optic effect. The shrinkage in diameter and decrease in RI change the ERIs of the two guided modes, and the prolongation of the waist modifies the coupling length. As a result, the phase difference ϕ experiences a variation and finally reflects the change in the output light intensity of P_3 and P_4. The tensile strain of the optical fiber is $\delta = \partial L/L$. Thus, the sensitivity of the corresponding coupler strain can be deduced as [16]:

$$S_\delta = \frac{\partial \lambda_N}{\partial \delta} = \frac{\lambda_N}{n_g^{even} - n_g^{odd}} \left(\Delta n_{\text{eff}} + \frac{\partial \Delta n_{\text{eff}}}{\partial \delta} \right) \tag{4}$$

where $\Delta n_{\text{eff}} = n_{\text{eff}}^{even} - n_{\text{eff}}^{odd}$, $n_g = n_{eff} - \lambda_N \partial n_{eff}/\partial \lambda$ represents the group ERI of the guided mode, and $G = n_g^{even} - n_g^{odd}$ denotes the group ERI difference between the two modes. It should be noted that the term $\partial(\Delta n_{\text{eff}})/\partial \delta$ is dependent on both the elasto-optic coefficient and the Poisson's ratio of fused silica. Equation (4) reveals that when $G = 0$, the sensitivity of the nanofiber coupler to axial strain can be improved significantly. The wavelength which satisfies $G = 0$ is also known as the dispersion turning point, and it can only be satisfied when the group ERIs of the two modes equal each other.

In order to gain a straightforward understanding of the sensing performance for the optical nanofiber coupler to axial strain and, hence, acoustic waves, we carried out numerical simulations. We first calculated the group ERI difference G between the odd mode and the even mode for a nanofiber coupler with a diameter of 1.6 μm, within the wavelength range of 1200~1700 nm. As shown in Figure 3a, the group ERI difference G varies from a negative value to a positive value as the working wavelength gradually increases from 1200 nm to 1660 nm, and it is equal to 0 at a wavelength of about 1482 nm. This critical wavelength is the DTP for the nanofiber coupler with a diameter of 1.6 μm. According to Equation (4), the axial strain sensitivity can reach infinity when λN is divided by 0.

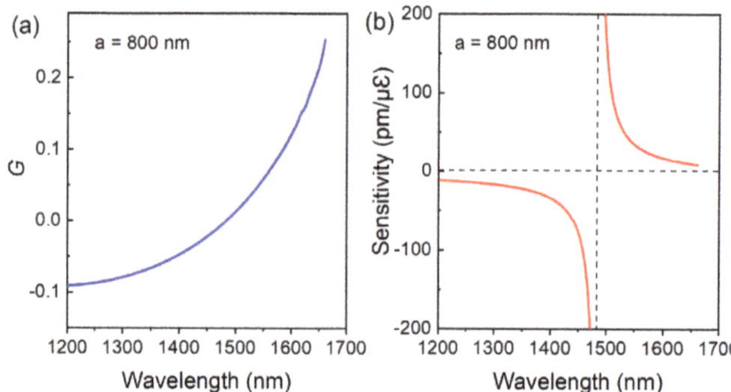

Figure 3. (**a**) Effective refractive index difference between the odd and even modes. (**b**) The strain sensitivity of the ONC.

Then, we calculated the axial strain sensitivity of the sensor according to Equation (4). The calculation results in Figure 3b reveal that the sensitivity curve shows a rectangular hyperbola shape. The axial strain sensitivity is significantly enhanced towards $-\infty$ on the left side of the DTP and towards $+\infty$ on the right side. This means that when interference dips/peaks approach the DTP, the wavelength shifts induced by axial strain can be greatly enhanced. According to our previous research, the DTP for nanofiber couplers in air can be tuned from 940 to 1670 nm simply by increasing the fiber diameter from 500 to 900 nm. By utilizing the DTP, ultrahigh sensitivity of nearly 100 nm/με can be achieved, which is promising for acoustic wave sensing and vibration measurement applications.

2.2. Working Principle of the Diaphragm

Due to the small diameter of the ONC, it is difficult for sound pressure to act on the ONC itself. So, a diaphragm is used as the sound-pressure-sensitive element to convert the acoustic wave to the mechanical vibration of itself. The ONC attached to the diaphragm can sense the acoustic wave indirectly by monitoring the vibration of the diaphragm. When the acoustic wave acts on the sensing diaphragm, the diaphragm will deform. When the pressure applied to the diaphragm is FP, the bending deformation of the diaphragm is [17,18]:

$$d = \frac{3F_P R^4 \left(1 - \mu^2\right)}{64 E h^3} \left(1 - \frac{r^2}{R^2}\right)^2 \tag{5}$$

where μ and E are the Poisson's ratio and Young's modulus of the diaphragm material, respectively. R and h denote the radius and thickness of the diaphragm, respectively, and r denotes the distance from a point on the diaphragm to the center of the diaphragm.

According to (5), the cross-sectional curve of the deformation for a circular diaphragm can be obtained, as shown in Figure 4. It can be seen that the maximum deformation occurs at the center of the diaphragm. This indicates that to obtain a higher acoustic sensitivity, one end of the ONC should be fixed at the center of the diaphragm, and the other end should be fixed at the edge. In this way, the optical fiber coupler has the maximum deformation when the acoustic wave acts on the diaphragm.

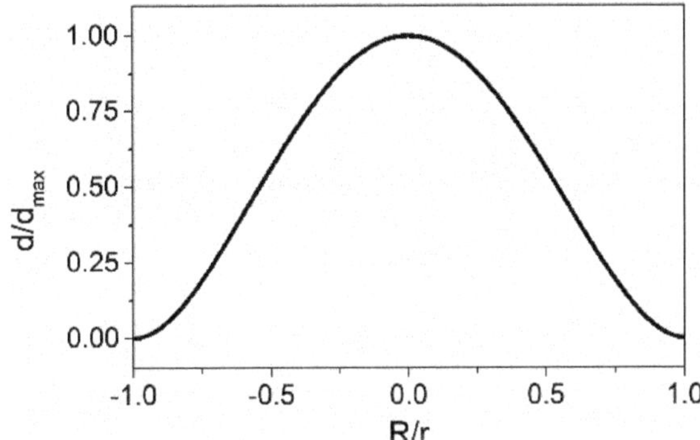

Figure 4. Normalized deformation at any point of the diaphragm, as a function of the normalized distance to the center.

Then, we analyzed the broadband acoustic response characteristics of the diaphragm. The selected diaphragm is a circular PE diaphragm with a radius of 4 cm and a thickness of 50 μm. The resonance frequency of the diaphragm is [17,18]:

$$f_{mn} = \frac{k_{mn}^2 h}{4\pi h R^2} \sqrt{\frac{E}{3\rho(1 - \mu^2)}} \tag{6}$$

where the k_{mn} value of the one-dimensional circular diaphragm is 3.196; R and h are the radius and thickness of the diaphragm, respectively. E and ρ are the Young's modulus and density of the diaphragm, respectively.

The dynamic deformation of the diaphragm is [17,18]:

$$\Delta d = \frac{3\left(1 - \mu^2\right) R^4 P}{16 E h^3} \frac{f_{mn}^2}{\sqrt{\left(f_{mn}^2 - f_a^2\right)^2 + 4 f_a^2 \beta^2}} \tag{7}$$

where β is the damping coefficient, f_a is the acoustic wave frequency, and P is the sound pressure. The calculation results are shown in Figure 5, which indicates that the diaphragm has the largest deformation when the acoustic wave is about 122.4 Hz, and the deformation is relatively gentle at 500 Hz–20 kHz.

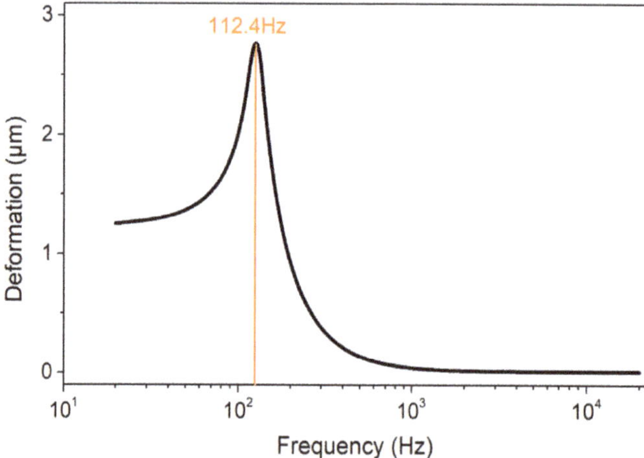

Figure 5. Diaphragm deformation diagram under different frequency acoustic waves when the pressure is 1 Pa.

3. Sensor Fabrication and Acoustic Measurement System

Optical nanofiber couplers were fabricated from standard telecommunication single-mode optical fibers by the fusion elongation method. Generally speaking, two sections of bare single-mode fibers were double twisted and fixed by two fiber clamps. Then, the fibers were heated to the glass transition temperature by an alcohol lamp, and two motorized translation stages stretched the fibers. The two optical fibers shrank in diameter and were fused together; finally, the nanofiber coupler was obtained. To ensure that the optical nanofiber couplers possessed a DTP at the desired wavelength, an online monitoring system was used to track the output spectrum of the fiber coupler in real time. The tapering process was terminated once the DTP appeared at the desired wavelength in the output spectrum. Figure 6 depicts a representative micrograph of a fabricated optical nanofiber coupler.

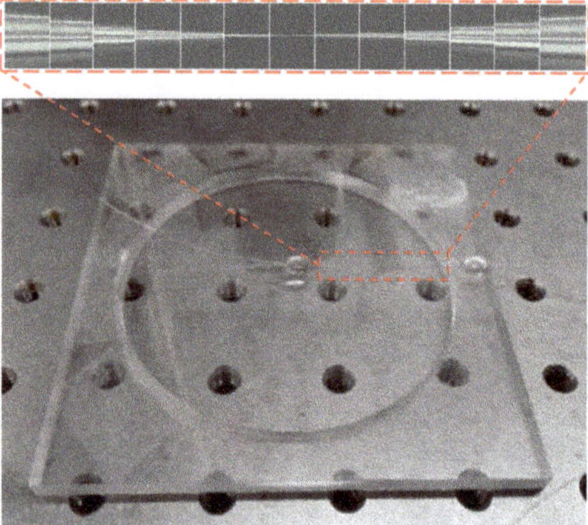

Figure 6. Photograph of the nanofiber coupler acoustic sensor.

The pedestal for the PE diaphragm was fabricated by a CNC machining center. The radius for the central hole was 4 cm. The PE film was fixed to the pedestal using UV glue, and a circular diaphragm with a radius of 4 cm was formed. Then, we fixed one end of the fabricated nanofiber coupler at around the center of the diaphragm and the other end at the edge with UV glue. A photograph of the fabricated acoustic wave sensor is shown in Figure 6.

The acoustic wave measurement system for the optical nanofiber coupler sensor is shown in Figure 7. The system consists of a sensing module and a signal demodulation module. The sensing part includes a tunable band laser (SANTEC TSL-550, Santec Corporation, Komaki, Aichi, Japan), the nanofiber coupler acoustic sensor, and two avalanche photodiodes (APDs, THORLABS PDA20C2, Thorlabs, Newton, NJ, USA). The signal demodulation part includes the filtering and amplifying circuits, a data acquisition card (DAQ, ART usb3133A, ART Technology, Beijing, China), and a computer. The light emitted by the laser enters one input arm of the nanofiber coupled acoustic wave sensor, and the optical signals from the two output arms are detected by the APDs. Then, after filtering, the data are collected by the DAQ, and finally, the signal is displayed, analyzed, and processed by the computer. Acoustic wave signals with the desired waveform and frequency were generated by a speaker driven by a function generator. A commercial sound pressure meter (SMART SENSOR AS824, Walfront LLC., Lewes, DE, USA) was used to measure the sound pressure of the acoustic wave signal.

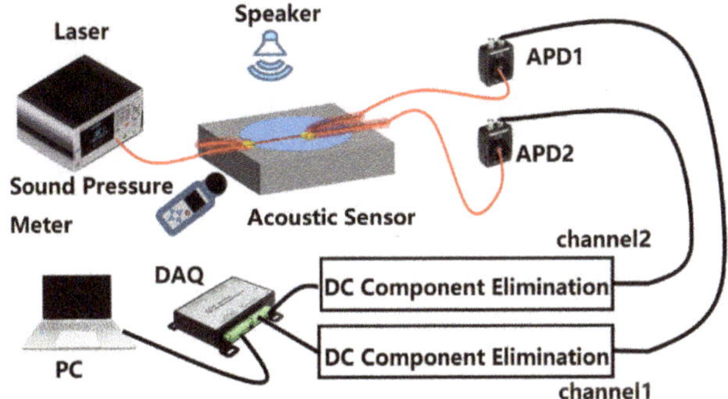

Figure 7. Schematic diagram of the acoustic wave measurement system.

4. Experimental Results and Discussion

The output spectrum from Port 3 of the ONC is displayed in Figure 8. It can be seen that the light wavelength corresponding to the DTP is about 1573 nm. Therefore, the output wavelength of the laser was set to 1573 nm, and this wavelength was chosen as the working wavelength of the ONC-based acoustic sensor. To improve the sensitivity for acoustic wave measurement, the difference between the output signals from Port 3 and Port 4 was used to measure acoustic waves. Figure 9 displays the detected signals from Ports 3 and 4 and their difference for acoustic signals of 40 Hz. It can be seen from the figure that the signal amplitude increased significantly after the difference operation.

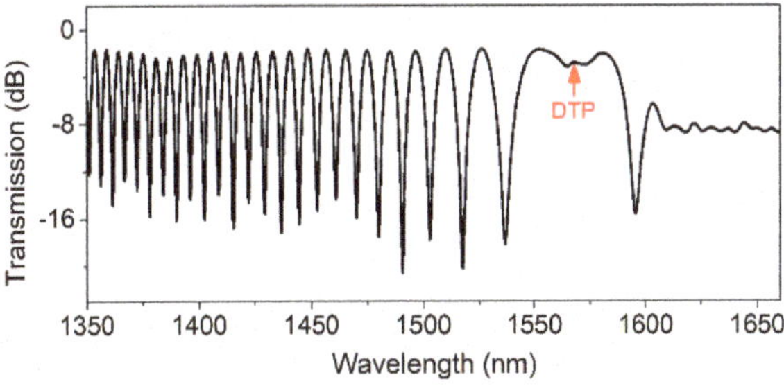

Figure 8. Spectrogram of the fabricated ONC with the DTP.

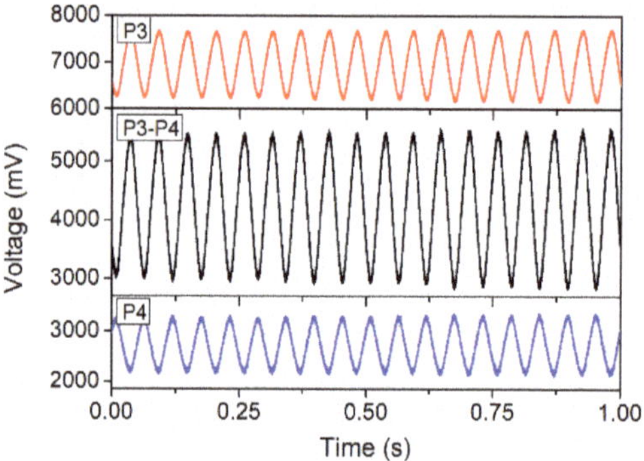

Figure 9. Measured signals from Ports 3 and 4 and the difference between Ports 3 and 4 under a 40 Hz acoustic wave.

4.1. Sensing Performance for Different Sound Pressures

First, we evaluated the performance of the nanofiber coupler acoustic sensor in response to different sound pressures. We applied an acoustic wave with a frequency of 160 Hz to the sensor and varied the loudness of the acoustic wave sequentially. The sound pressure was measured using a sound pressure meter. The output signal from the acoustic sensor is summarized in Figure 10, proving that the output voltage of the sensor acoustic signal is linear with the input sound pressure.

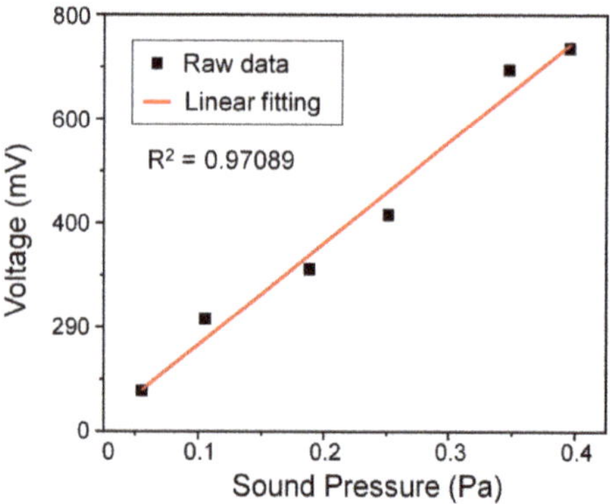

Figure 10. Linear diagram of output voltage under different sound pressures.

4.2. Broadband Acoustic Detection

Then, we analyzed the broadband acoustic detection capability of the nanofiber coupler acoustic sensor. We loaded acoustic wave signals with frequencies ranging from 30 Hz to 20 kHz on the sensor diaphragm and obtained the time-domain output signal using the demodulation system. We performed an FFT transformation on the time-domain signals to obtain the frequency-domain diagram of the different signals.

The proposed optical-nanofiber-coupler-based acoustic wave sensor shows a good response to acoustic waves in the broad frequency range of 30 Hz–20 kHz. The representative measured time-domain signals and frequency-domain diagrams for acoustic waves with frequencies of 30 Hz, 600 Hz, 3000 Hz, and 20,000 Hz are shown in Figure 11. The results reveal that our sensor shows a high signal-to-noise ratio in this wide frequency range.

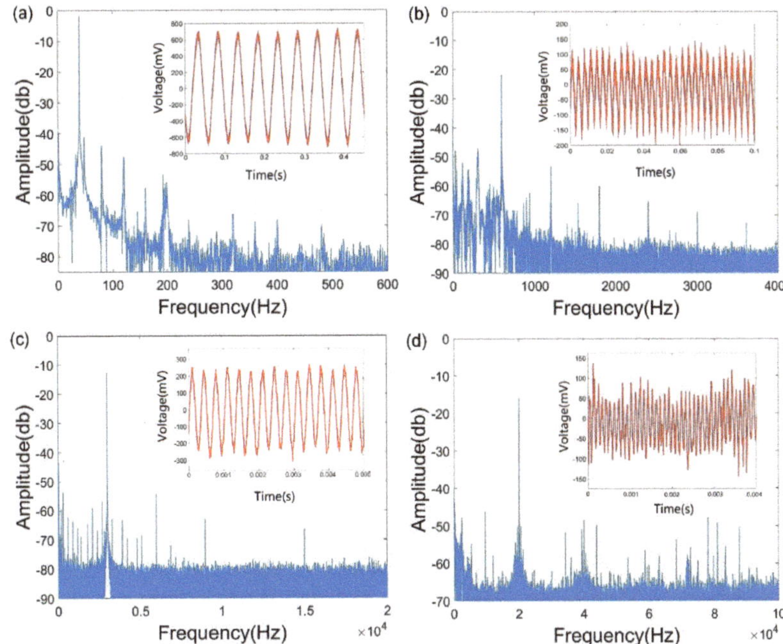

Figure 11. (**a–d**) Time-domain and frequency-domain diagrams of the sensor at frequencies of 30 Hz, 600 Hz, 3000 Hz, and 20,000 Hz, respectively.

The sensitivity of our sensor in this broadband frequency range was calculated based on the experimental results. As a comparison, we also tested the performance of our sensor with working wavelengths of 1500 nm and 1600 nm, respectively. One of these two working wavelengths lies on the left side of the DTP, and the other lies on the right side. According to the results of our numerical investigation, the optical nanofiber coupler sensor would show inferior performance working at these two wavelengths as compared to working at the DTP. The sensing performance in the frequency range of 30 Hz–20 kHz was measured using the demodulation system, and the sensitivities were also calculated.

All the results are displayed in Figure 12. After comparing and analyzing the respective sensitivities of the sensor working at 1500 nm, 1573 nm, and 1600 nm, it was found that the sensor showed higher sensitivities in the measured frequency range when working at the DTP than when working at the other two wavelengths. These results are consistent with our numerical results. Also, the sensor showed relatively high sensitivities in the low-frequency range of 30~500 Hz. The highest sensitivity was discovered at the frequency of 120 Hz, which agrees well with our numerical simulation result of 112.4 Hz. The small discrepancy may be induced by the UV glue and the optical nanofiber coupler that is fixed at the center of the diaphragm. The sensitivity change was relatively stable from 500 Hz to 20 kHz. Currently, the sensitivity is between 8 and 30 mv/Pa without the amplifier circuit.

We further analyzed the sensing performance of the sensor when a 120 Hz acoustic wave signal was loaded, as the sensor showed the highest sensitivity at this frequency. The frequency spectrum obtained through FFT transformation is shown in Figure 13. The prominent peak at 120 Hz corresponds to the frequency of the applied acoustic signal. The signal-to-noise ratio of the sensor is 38.4 dB, the minimum detectable sound pressure is 330 μPa/Hz$^{1/2}$, and the sensitivity is 1923 mV/Pa. Harmonic signals also appeared in the frequency spectrum, which may be caused by the reflection of acoustic waves in the cavity of the sensor base between the diaphragm and the test bench. These harmonic signals are negligible compared to the fundamental frequency signal. Our study demonstrates that optical nanofiber couplers working at the DTPs are promising candidates for developing high-sensitivity acoustic wave sensors.

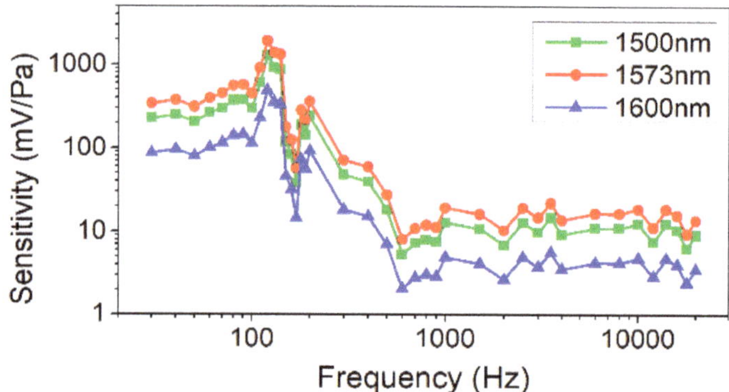

Figure 12. Sensitivity curves at different acoustic frequencies.

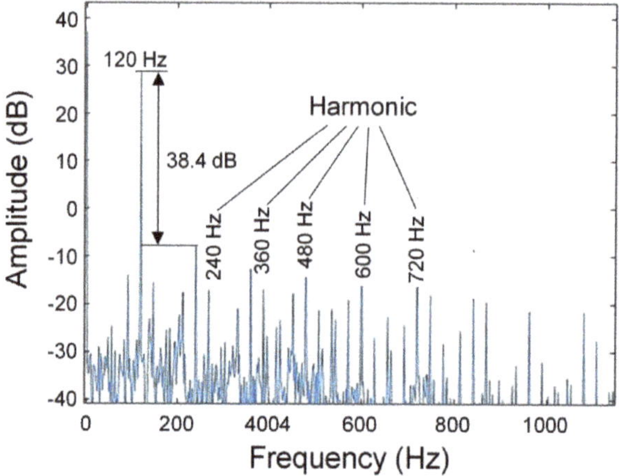

Figure 13. Frequency-domain diagram of the measured acoustic signal (operation wavelength: 1573 nm, acoustic wave frequency: 120 Hz).

5. Conclusions

In summary, in this paper, we proposed an ultrasensitive acoustic wave sensor based on a micro/nanofiber coupler operating at the dispersion turning point and a sensing diaphragm. The sensor was realized by fixing the fiber coupler on the diaphragm and converting the vibration of the diaphragm to the stretching of the fiber coupler. Our theoretical studies showed that the sensing performance can be improved significantly by utilizing the DTP, and the diaphragm can provide a broadband response to acoustic waves. Guided by our theoretical findings, we experimentally realized ultra-high sensitivity and broadband detection of acoustic waves using a nanofiber coupler working at the DTP and a sensing diaphragm. The sensor shows high sensitivity in low-frequency measurement and flat response to medium- and high-frequency acoustic signals, with the highest sensitivity of 1929 mV/Pa achieved at 120 Hz. This sensor has the advantages of being simple in construction and easy for demodulation, and it may have potential applications in seismic wave detection.

Author Contributions: Writing—original draft, X.G. and J.W. (Jiajie Wen); Writing—review & editing, X.G., J.W. (Jiajia Wang) and K.L. All authors have read and agreed to the published version of the manuscript.

Funding: This project was supported in part by the National Natural Science Foundation of China (No. 62005101, and No. 51905155); in part by the Key R & D of Jilin Provincial Department of Science and Technology-key Industrial Core Technology Research Project (20210201091GX); in part by the Basic and Applied Basic Research Foundation of Guangdong Province (No. 2022A1515010235); and in part by the State Key Laboratory of Applied Optics (NO.SKLAO2021001A03).

Institutional Review Board Statement: Not applicable.

Informed Consent Statement: Not applicable.

Data Availability Statement: The datasets are available from the corresponding author on reasonable request.

Conflicts of Interest: The authors declare no conflict of interest.

References

1. Teixeira, J.G.V.; Leite, I.T.; Silva, S.; Frazão, O. Advanced fiber-optic acoustic sensors. *Photonic Sens.* **2014**, *4*, 198–208. [CrossRef]
2. Jo, W.; Kilic, O.; Digonnet, M.J.F. Highly Sensitive Phase-Front-Modulation Fiber Acoustic Sensor. *J. Light. Technol.* **2015**, *33*, 4377–4383. [CrossRef]
3. Liu, L.; Lu, P.; Wang, S.; Fu, X.; Sun, Y.; Liu, D.; Zhang, J.; Xu, H.; Yao, Q. UV Adhesive Diaphragm-Based FPI Sensor for Very-Low-Frequency Acoustic Sensing. *IEEE Photonics J.* **2015**, *8*, 1–9. [CrossRef]
4. Diaphragm, M.G.; Ma, J.; Xuan, H.; Ho, H.L.; Jin, W.; Yang, Y.; Fan, S. 2013 Fiber-Optic Fabry–Perot Acoustic Sensor with Multilayer Graphene Diaphragm. *IEEE Photonics Technol. Lett.* **2013**, *25*, 932–935.
5. Wu, Y.; Yu, C.; Wu, F.; Li, C.; Zhou, J.; Gong, Y.; Rao, Y.; Chen, Y. A Highly Sensitive Fiber-Optic Microphone Based on Graphene Oxide Membrane. *J. Light. Technol.* **2017**, *35*, 4344–4349. [CrossRef]
6. Chen, L.H.; Chan, C.C.; Yuan, W.; Goh, S.K.; Sun, J. High performance chitosan diaphragm-based fiber-optic acoustic sensor. *Sens. Actuators A Phys.* **2010**, *163*, 42–47. [CrossRef]
7. Guo, F.; Fink, T.; Han, M.; Koester, L.; Turner, J.; Huang, J. High-sensitivity, high-frequency extrinsic Fabry–Perot interferometric fiber-tip sensor based on a thin silver diaphragm. *Opt. Lett.* **2012**, *37*, 1505–1507. [CrossRef] [PubMed]
8. Dass, S.; Jha, R. Tapered Fiber Attached Nitrile Diaphragm-Based Acoustic Sensor. *J. Light. Technol.* **2017**, *35*, 5411–5417. [CrossRef]
9. Wang, S.; Lu, P.; Zhang, L.; Liu, D.; Zhang, J. Optical Fiber Acoustic Sensor Based on Nonstandard Fused Coupler and Aluminum Foil. *IEEE Sens. J.* **2014**, *14*, 2293–2298. [CrossRef]
10. Ni, W.; Lu, P.; Fu, X.; Wang, S.; Sun, Y.; Liu, D.; Zhang, J. Highly Sensitive Optical Fiber Curvature and Acoustic Sensor Based on Thin Core Ultralong Period Fiber Grating. *IEEE Photonic J.* **2017**, *9*, 1–9. [CrossRef]
11. Liu, K.; Fan, J.; Luo, B.-B.; Zou, X.; Wu, D.; Zou, X.; Shi, S.; Guo, Y.; Zhao, M. Highly sensitive vibration sensor based on the dispersion turning point microfiber Mach-Zehnder interferometer. *Opt. Express* **2021**, *29*, 32983. [CrossRef] [PubMed]
12. Li, K.; Zhang, T.; Liu, G.; Zhang, N.; Zhang, M.; Wei, L. Ultrasensitive optical microfiber coupler based sensors operating near the turning point of effective group index difference. *Appl. Phys. Lett.* **2016**, *109*, 101101. [CrossRef]
13. Li, K.; Zhang, N.M.Y.; Zhang, N.; Zhang, T.; Liu, G.; Wei, L. Spectral Characteristics and Ultrahigh Sensitivities Near the Dispersion Turning Point of Optical Microfiber Couplers. *J. Light. Technol.* **2018**, *36*, 2409–2415. [CrossRef]
14. Li, K.; Zhang, N.; Zhang, N.M.Y.; Liu, G.; Zhang, T.; Wei, L. Ultrasensitive measurement of gas refractive index using an optical nanofiber coupler. *Opt. Lett.* **2018**, *43*, 679–682. [CrossRef] [PubMed]
15. Zhou, W.; Li, K.; Wei, Y.; Hao, P.; Chi, M.; Liu, Y.; Wu, Y. Ultrasensitive label-free optical microfiber coupler biosensor for detection of cardiac troponin I based on interference turning point effect. *Biosens. Bioelectron.* **2018**, *106*, 99–104. [CrossRef] [PubMed]
16. Wen, J.; Yan, X.; Gao, X.; Li, K.; Wang, J. Axial Strain Sensor Based on Microfiber Couplers Operating at the Dispersion Turning Point. *IEEE Sens. J.* **2022**, *22*, 4090–4095. [CrossRef]
17. Aiadi, K.E.; Rehouma, F.; Bouanane, R. Theoretical analysis of the membrane parameters of the fiber optic microphone. *J. Mater. Sci. Mater. Electron.* **2006**, *17*, 293–295. [CrossRef]
18. Buchade, P.B.; Shaligram, A.D. Simulation and experimental studies of inclined two fiber displacement sensor. *Sens. Actuators A Phys.* **2006**, *128*, 312–316. [CrossRef]

Article

An Ultra-High-Resolution Bending Temperature Decoupled Measurement Sensor Based on a Novel Core Refractive Index-like Linear Distribution Doped Fiber

Yunshan Zhang [1,*], Yulin Zhang [1], Xiafen Hu [2], Dan Wu [2], Li Fan [3], Zhaokui Wang [1] and Linxing Kong [4,*]

1   School of Aerospace Engineering, Tsinghua University, Beijing 100084, China;
     y.l.zhang@mail.tsinghua.edu.cn (Y.Z.); wangzk@tsinghua.edu.cn (Z.W.)
2   System Design Institute of Hubei Aerospace Technology Academy, Wuhan 430040, China;
     xiafenhu@163.com (X.H.); wudanbd@163.com (D.W.)
3   Beijing Tianji Space Technology Co., Ltd., Beijing 100084, China; fanli77@mail.tsinghua.edu.cn
4   School of Opto-Electronic Information Science and Technology, Yantai University, Yantai 264005, China
*   Correspondence: yunshanzhangedu@163.com (Y.Z.); 1120170096@mail.nankai.edu.cn (L.K.)

Abstract: A high-resolution and high-sensitivity fiber optic sensor based on the quasi-linear distribution of the core refractive index is designed and fabricated, which enables decouple measurement of bending and of temperature. First, single-mode fiber doped with Al_2O_3, Y_2O_3, and P_2O_5 was drawn through a fiber drawing tower. The fiber grating was engraved on the fiber by a femtosecond laser. Modeling analysis was conducted from quantum theory. Experimental results show that the bending sensitivity of the grating can reach 21.85 dB/m^{-1}, which is larger than the reported sensitivity of similar sensors. In the high temperature range from room temperature to 1000 °C, the temperature sensitivity was 14.1 pm/°C. The doped grating sensor can achieve high temperature measurement without annealing, and it has a distinguished linear response from low temperature to high temperature. The bending resolution can reach 0.0004 m^{-1}, and the temperature resolution can reach 0.007 °C. Two-parameter decoupling measurement can be realized according to the distinctive characteristic trends of the spectrum. What's more, the sensor exhibits excellent stability and a fast response time.

Keywords: high resolution; bending sensor; temperature sensor; decoupling measurement

Citation: Zhang, Y.; Zhang, Y.; Hu, X.; Wu, D.; Fan, L.; Wang, Z.; Kong, L. An Ultra-High-Resolution Bending Temperature Decoupled Measurement Sensor Based on a Novel Core Refractive Index-Like Linear Distribution Doped Fiber. *Sensors* **2022**, *22*, 3007. https://doi.org/10.3390/s22083007

Academic Editor: Jin Li

Received: 8 March 2022
Accepted: 11 April 2022
Published: 14 April 2022

Publisher's Note: MDPI stays neutral with regard to jurisdictional claims in published maps and institutional affiliations.

1. Introduction

With corresponding concepts such as the Internet of Things and smart cities, the sensing of information is an important content of the information age. As the perception layer of the information age, sensors are the entrance to the reception of massive amounts of information, and they are an essential foundation for the internet of everything. Sensing technology is evolving towards miniaturization, intelligence, integration, and passivity. The scope of physical perception is required to be more extensive, the means of information collection are more convenient, and the types of data acquisition are more diverse.

The two physical parameters of temperature and curvature need to be evaluated in many applications, such as aerospace, petroleum energy, smart wearables, and industrial manufacturing [1]. After years of research, fiber bending and temperature sensors based on different principles have been reported. For example, the Mach–Zehnder interferometer [2], photonic crystal fiber [PCF] [3], four-core sapphire-derived fiber [4], seven core fiber [5], optical fiber laser [6], two-core fiber [7], three-core fiber [8], fiber Fabry–Perot cavity [9], Michelson interferometer [10], Sagnac Interferometer [11], long period fiber grating (LPFG) [12], fiber Bragg grating (FBG) [13], and capillary hollow-core fiber [14]. These sensors have their own unique advantages, but there are still many shortcomings. Interferometric sensors need to be fused multiple times, resulting in the reduced mechanical strength of the sensor [15,16]. Multi-core fiber and PCF are expensive to use [17]. Although the LPFG sensor has high sensitivity, it usually analyzes the data through a transmission spectrum, and it is therefore difficult to reuse the optical fiber sensor network [18]. When an FBG sensor is used to evaluate bending and temperature, it has low sensitivity and it cannot distinguish measurement [19]. In terms of temperature measurement, the sapphire grating sensor can measure very high temperatures. However, the cost is high and the manufacturing technology is not mature enough [20]. The regenerative grating can measure high temperature, but it requires a long period of high temperature annealing treatment, after which the grating area will be very brittle [21].

Therefore, in order to address the above problems, we propose a FBG sensor engraved by femtosecond laser on a new doped fiber. First, we designed and drew a new type of fiber doped with Al_2O_3, Y_2O_3, and P_2O_5. The core refractive index of the new optical fiber is quasi-linear distribution. The FBG is engraved on the fiber by the femtosecond laser. The grating model is analyzed by quantum optics theory and refractive index perturbation theory. Bending and temperature experiments were performed on this new type of grating. Experimental results found that the grating exhibited very high bending sensitivity. The grating can achieve continuous temperature measurement without long-term high temperature annealing. The grating spectral energy is sensitive to bending, and the wavelength is sensitive to temperature. It exhibits unique spectral properties for two different physical quantities. The sensor measurement system has ultra-high curvature and temperature resolution.

2. Fabrication and Theoretical Analysis

The purpose of traditional optical fiber doping materials is to obtain the refractive index difference between the core and the cladding. However, the purpose of doped oxide materials in this manuscript is to enhance the temperature resistance properties of the fiber, while modulating the refractive index distribution of the core. The melting point of Al_2O_3 is 2054 °C. Additionally, Al_2O_3 has the characteristics of a small specific surface area, uniform particle size, easy dispersion, high hardness and good insulation performance, which makes the fabrication of optical fiber preform and the drawing of optical fiber easier to control. Meanwhile, Al_2O_3 has the characteristics of high mechanical strength, strong wear resistance, and heat shock resistance, which makes the sensor more robust. It holds the characteristics of colorless and transparent, high light transmittance and high refractive index [22]. The melting point of Y_2O_3 is 2439 °C. Below 2200 °C, Y_2O_3 is a cubic phase without birefringence [23,24]. It holds the characteristics of corrosion resistance, wide optical transparency, and good physicochemical properties.

A new type of optical fiber can be obtained by drawing the optical fiber preform through the optical fiber drawing tower. The high-temperature resistance furnace of the drawing tower is heated to 2300 °C, so that the cone of the optical fiber preform is melted to form a droplet like material head. After controlling the thinning of the optical fiber preform, it goes through an annealing device and a cooling system in turn. The dimension parameters of optical fibers are supervised online in real time. Finally, the bare fiber is coated and cured by means of a coating device. Figure 1 presents a schematic diagram to reinforce the description of the novel fiber fabrication.

Table 1 presents the measured dimensional parameters and numerical apertures of the novel fiber.

The doped optical fiber core was analyzed by electron probe micro analysis (EPMA). The EPMA enables qualitative and quantitative analysis of chemical composition of microscopic regions in fiber. The mass percentage of different doping can be achieved by fully automatic area scanning analysis. Its core contains 4.9 wt.%Y_2O_3, 3.2 wt.%Al_2O_3, and 2.3 wt.% P_2O_5. It is the most sensitive optical fiber measurement technology available in the testing wavelength range of 375 nm~2000 nm, which can measure the refractive index, stress, and geometry of any type of fiber. Table 2 gives the detailed parameters of FRIMA-IFA-100-IR.

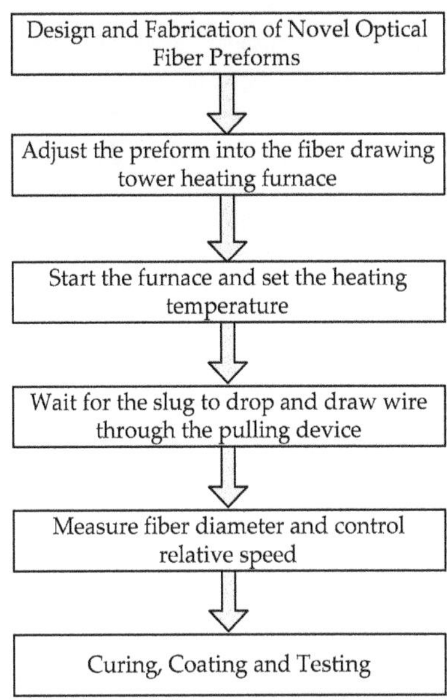

Figure 1. Schematic diagram of new optical fiber manufacturing.

Table 1. Parameters of novel high temperature resistant doped fiber.

Core Composition	Cladding Diameter	Core Diameter	Coating Diameter	Numerical Aperture
SiO_2-Al_2O_3-Y_2O_3-P_2O_5	125 μm	10 μm	252 μm	0.2

Table 2. Measured parameters of FRIMA-IFA-100-IR.

Refractive Index Measurement Accuracy	Spatial Resolution	Measuring Concentricity Error	Measure Core Out-of-Roundness Error
±0.0001	500 nm	±200 nm	±0.4%

Refractive index measurements of fiber cross-sections are set out in Figure 2. The maximum refractive index is 1.464, and the cladding refractive index is significantly less than the core index. The abscissa and the ordinate represent the number of measurement sampling points in Figure 2a. Locally amplified refractive index profile measurements were performed on the fiber core region and the results are presented in Figure 2b. The point of maximum refractive index of the fiber is located in the center of the fiber core. Its corresponding core refractive index presents a non-uniform distribution of the point spread function. The refractive index difference in the horizontal direction and in the vertical direction of the fiber were measured, and the results are depicted in Figure 2c. From the results, it is found that the refractive index distribution of the fiber cladding is relatively uniform, but the core refractive index almost implies a linear decrease. In addition to this, the core has a small amount of jitter in the horizontal and in the vertical refractive index profiles. When light is transmitted in the new optical fiber structure, the mode of the light field will change, causing the total reflection angle of light to become smaller when light is transmitted in the optical fiber. When the optical fiber is bent, it will cause part of the optical energy to be coupled into the optical fiber cladding. Taking advantage of this property of the new type of optical fiber, the sensor can be used for bending measurements by demodulating the energy.

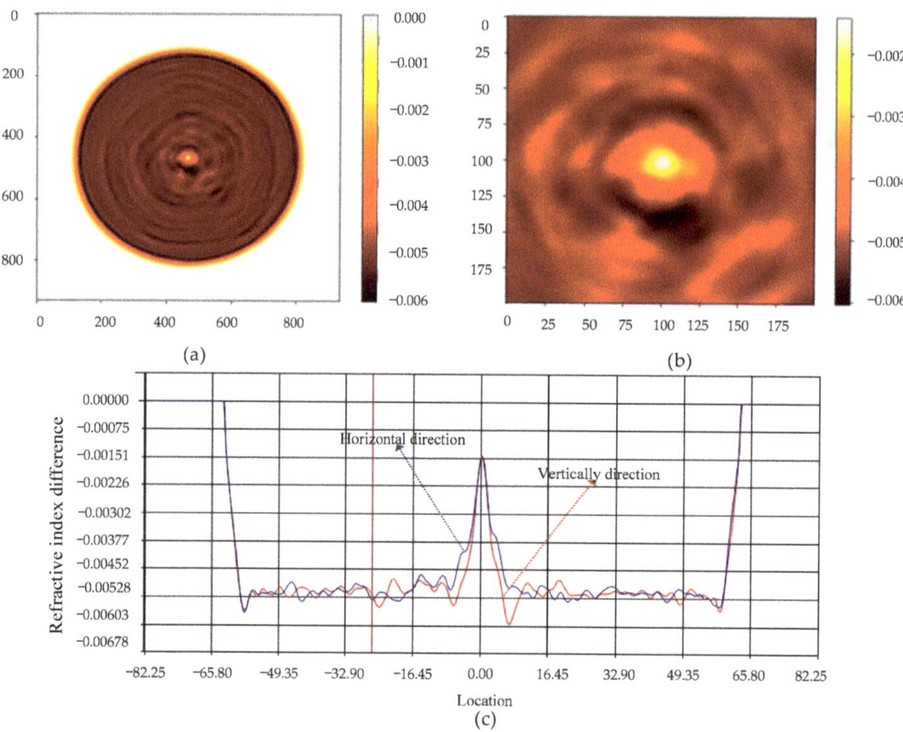

Figure 2. New fiber parameters, (**a**) Optical fiber cross-section refractive index profile; (**b**) Fiber Core Refractive Index Profile; (**c**) Refractive index difference in horizontal direction and in vertical direction of optical fiber.

Fiber gratings were engraved by direct writing using the femtosecond laser micro-machining system. Femtosecond lasers are characterized by ultra-short pulse widths and ultra-high peak powers. When the femtosecond laser interacts with the fiber material, non-linear ionization phenomena such as nonlinear photoionization and avalanche ionization are mainly generated. The high-energy femtosecond laser acts on the fiber to cause the formation of internal defects or local shrinkage of the material. It densifies the material and causes localized melting of the fiber core, resulting in refractive index modulation for permanent material damage. What's more, the avalanche ionization effect is generated and the fiber material continuously absorbs the laser irradiation energy. Eventually, the electron plasma in the laser focus area will increase to a certain concentration value, and more intense light absorption will occur, resulting in permanent refractive index modulation in the irradiated area.

The center wavelength of the femtosecond laser was 800 nm. The repetition rate was 200 kHz. The pulse width was 200 fs, and the energy range was tunable from 0 to 3 W. A laser energy attenuation system was used to adjust the laser output energy, and finally the refractive index was modulated on the new optical fiber by focusing on a high-magnification microscope objective. The processing position and the modulation effect of the front end of the laser imaged and observed by CCD. In the process of preparing the grating, the optical fiber was fixed on the three-dimensional displacement stage through the adsorption clamp platform. The computer terminal controlled the three-dimensional stage to make the fiber move at a specific speed and in a specific direction. The reflection spectrum appeared in real time through the demodulator. Figure 3 is the grating spectrum engraved under the above femtosecond laser parameters. It can be seen from Figure 2 that the grating reflection spectrum had a higher contrast and a lower noise. The center wavelength of the sample A grating was 1533.58 nm. The grating period was 1.2 μm. The number of lines engraved by the laser was 5000, and the corresponding grating length was 6 mm. The center wavelength of the sample B grating was 1533.58 nm. The grating period was 1.2 μm. The number of lines engraved by the laser was 3000, and the corresponding grating length was 3.6 mm. Figure 3a,b show the reflection spectra of two different grating samples with lengths of 6 mm and 3.6 mm, respectively. The length of the sample A grating was greater than that of the sample B grating. The period of the sample grating was the same. The difference was that the length of engraving was different under the same period.

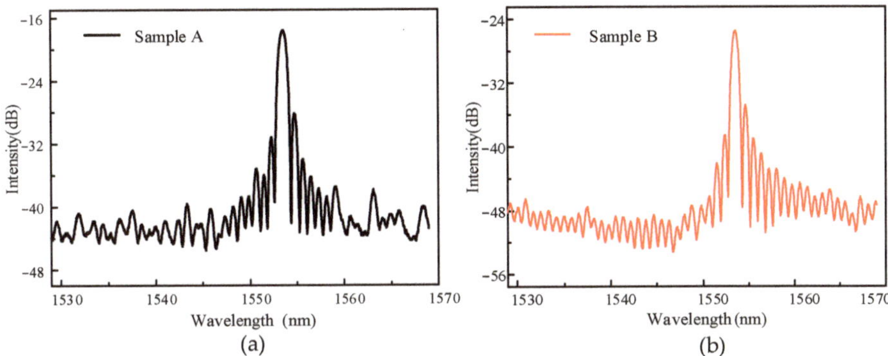

Figure 3. Reflection spectra of samples with different gratings. (**a**) Sample A; (**b**) Sample B.

This manuscript presents an analogous approach to theoretical analysis in which the micromachining of optical fibers by femtosecond lasers is considered as a perturbation of the refractive index of the fiber core. It is proposed to deduce the theory of grating coupled mode by establishing the perturbation theory applicable to Maxwell's equations. The main idea of modeling is as following: introduce the state vector

$$|\psi\rangle = \begin{pmatrix} \vec{E} \\ j\vec{H} \end{pmatrix} \tag{1}$$

and assume that the electromagnetic field changes with time in a relationship of $e^{j\omega t}$. So, Maxwell's equation can be expressed as:

$$\hat{L}_t|\psi\rangle + \hat{L}_z|\psi\rangle - \frac{\omega}{c}\hat{W}|\psi\rangle = 0 \tag{2}$$

where

$$\hat{L}_t = \begin{pmatrix} 0 & \nabla_t \times \\ \nabla_t \times & 0 \end{pmatrix}, \hat{L}_z = \begin{pmatrix} 0 & \frac{\partial}{\partial z}\vec{e}_z \times \\ \frac{\partial}{\partial z}\vec{e}_z \times & 0 \end{pmatrix}, \hat{W} = \begin{pmatrix} \overset{\leftrightarrow}{\varepsilon} & 0 \\ 0 & \overset{\leftrightarrow}{\mu} \end{pmatrix} \tag{3}$$

For an unperturbed fiber, the state vector corresponding to the propagation constant analysis β_k is $|\psi\rangle_k = e^{j\beta_k z}|\psi_k(x,y)\rangle$. $|\psi_k(x,y)\rangle$ is the transverse eigenfunction of the state vector, which satisfies the eigen equation

$$\left(-\hat{L}_t + \frac{\omega}{c}\hat{W}_0\right)|\psi_k\rangle = \beta_k\hat{\Gamma}_z|\psi_k\rangle \tag{4}$$

$$\hat{\Gamma}_z = j\begin{pmatrix} 0 & \vec{e}_z \times \\ \vec{e}_z \times & 0 \end{pmatrix}, \hat{W}_0 = \begin{pmatrix} \varepsilon_u & 0 \\ 0 & 1 \end{pmatrix} \tag{5}$$

when the optical fiber core is perturbed by the femtosecond laser, the perturbation form is set as

$$\hat{W}_\delta = \hat{W} - \hat{W}_0 \tag{6}$$

the electromagnetic field propagating in the fiber is the superposition state of the eigenmodes in the undisturbed fiber, namely

$$|\psi\rangle = \sum_k a_k(z)e^{j\beta_k z}|\psi_k\rangle \tag{7}$$

According to the time-dependent perturbation theory of quantum mechanics, the superposition state $|\psi\rangle$ can evolve according to the following equation

$$\hat{L}_z|\psi\rangle = \left(-\hat{L}_t + \frac{\omega}{c}\hat{W}_0\right)|\psi\rangle + \frac{\omega}{c}\hat{W}_\delta|\psi\rangle \tag{8}$$

Using the orthonormalization condition, the coupled mode equation for the mode amplitude can be obtained as follows

$$\frac{\partial}{\partial z}a_j(z) = j\sum_k a_k(z)e^{j(\beta_k - \beta_j)z}\langle\psi_j|\hat{W}_\delta|\psi_k\rangle \tag{9}$$

where $\hat{W}_\delta$ is related to the perturbation of the dielectric constant. If the perturbation form of the dielectric constant caused by the refractive index modulation is found, the corresponding perturbation coupled mode equation can be obtained.

3. Experiment and Discussion

Figure 4 is a curvature experiment measurement device system. One end of the grating was connected to a high-precision demodulator (FAZT), and its wavelength resolution was 0.1 pm. The demodulator is connected to the computer, and the reflection spectrum information of the grating can be presented in real time on the computer's software. The grating bending region is encapsulated in elastic capillaries to prevent the grating from breaking due to stress concentration. The elastic capillary is fixed on the two three-dimensional translation stages (TDSs) by the fiber holder, and the curvature of the sensor can be modified by changing the distance between the two translation stages. The grating area is always centered on both stages while preventing twisting of the sensor. When the fiber grating region is bent, its curvature function can be expressed as

$$\sin\frac{L}{2R} = \frac{D}{R} \tag{10}$$

where R is the radius of curvature of the fiber grating; L represents the initial distance between the two TDSs; and D is the absolute value of the travel distance between the TDSs.

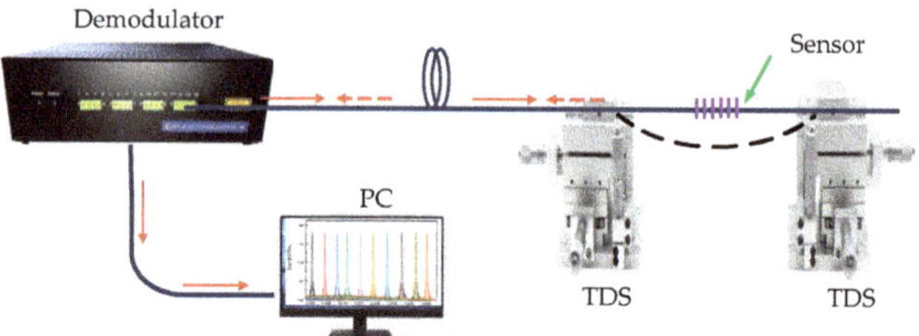

Figure 4. Bending experiment test device system.

Bending experiments were carried out on the novel grating sample A, and the results are shown in Figure 5a. The center wavelength of the sample A grating was 1533.58 nm. The grating period was 1.2 µm. The number of lines engraved by the laser was 5000, and the corresponding grating length was 6 mm. Ambient temperature remained constant when measuring curvature. The reflection spectra under different curvature values were recorded. The experimental results clearly reveal that the reflection spectral response of the grating was very sensitive with the change of curvature. Further analysis according to the changing characteristics of the spectrum shows that the energy of the resonant spectrum changes very rapidly to bending, but the wavelength is hardly affected. Through the experimental consequences, the potential trend of central spectral energy with curvature is analyzed again. The energy of the reflected spectrum drops sharply with increasing curvature. This is attributed to the fact that most of the light transmitted in the grating region are coupled into the fiber cladding with increasing curvature. The energy of the sensor changed by 27.5 dB when the curvature range increased from 0 to 1.2 m^{-1}. A linear fit was performed on a large number of data points, and the fitting results are shown in Figure 5b. The bending sensitivity of the grating was -21.85 dB/m^{-1} and the linearity amounted to 0.994, which is very convenient for signal demodulation. Since the energy resolution of the sensor signal demodulator was 0.01 dB, the bending resolution of the sensor can be obtained by calculation to be 0.0004 m^{-1}. A large number of experimental measurements were carried out, all exhibiting similar sensing properties. Therefore, in the following figures we give the specific experimental results of sample A to represent the sensing characteristics of this type of sensor.

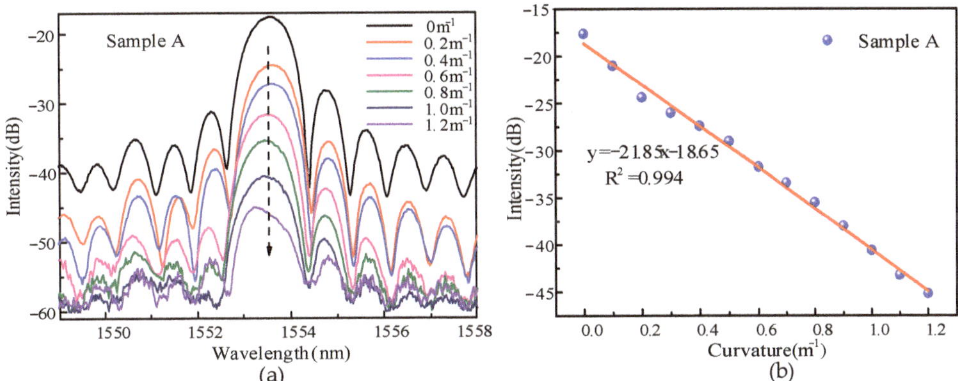

Figure 5. (**a**) Reflection spectrum of sensor under different curvature; (**b**) Bending fitting curve.

The stability of the sensor was checked for ten hours when the curvature was 0.4 m^{-1} and 1.0 m^{-1}, respectively. The experimental results are shown in Figure 6. The sensor demonstrates exceptional long-term stability. In the long-time measurement process, the maximum fluctuation of reflected spectral energy was 0.3 dB. Considering bending sensitivity of the grating, the curvature fluctuation of the bending was 0.014 m^{-1}.

The temperature characteristics of fiber gratings doped with high temperature resistant materials were checked. The heating furnace adopted high-quality high alumina polycrystalline ceramic fiber, which holds the characteristics of rapid heating resistance, stability, and uniform heating in the heating area. The maximum working temperature of the tubular furnace was 1100 °C, and the temperature control accuracy was 1 °C. The length of the heating zone was 300 mm. The temperature measurement area was calibrated by a soaring temperature thermocouple contact probe. Then, place the grating area in a free straightening state in a heating furnace and raise the temperature from room temperature to 1000 °C, and the room temperature is 20 °C; record the spectral data every 100 °C, and stabilize the temperature at the recording point for half an hour to heat it evenly. Figure 7 records the spectrum of grating at different temperatures. In Figure 6, we still give the experimental measurement results of sample A. The sensing properties of this novel sensor are represented by sample A.

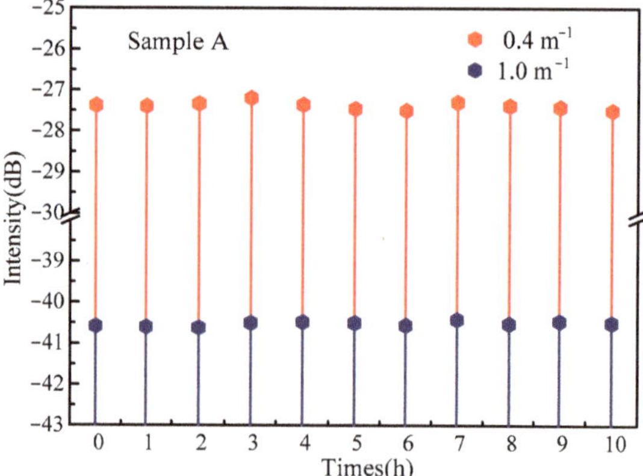

Figure 6. Bending stability test of sensor.

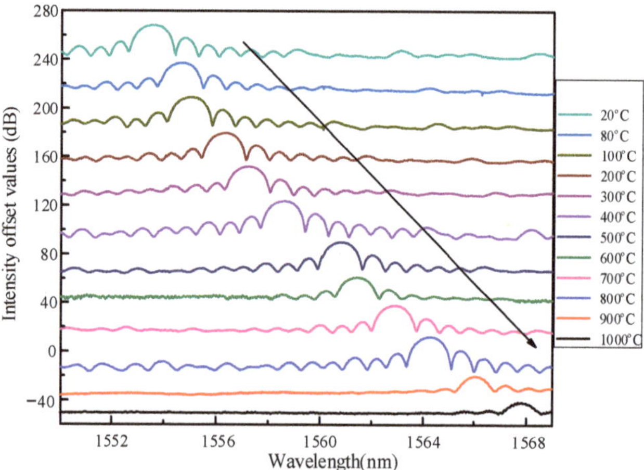

Figure 7. Grating reflection spectrum at different temperatures.

With the increase of temperature, the wavelength of the grating drifts in the long wave direction. The effect of temperature on the wavelength of fiber grating includes three aspects. The thermal expansion effect of the fiber causes the grating grid spacing to change. The optical fiber thermo-optic effect changes the refractive index of the fiber grating. The elastic-optic effect caused by the thermal stress inside the fiber makes the core diameter of the fiber change. The total effect of temperature on FBG wavelength drift is:

$$\frac{\Delta\lambda_B}{\lambda_B} = (\alpha + \zeta)\Delta T \tag{11}$$

where λ_B is the center wavelength of the grating; α is the thermal expansion coefficient of the fiber material, and its value is greater than zero; and ζ is the thermo-optic coefficient of the fiber material, and its value is greater than zero. So, $\Delta\lambda_B$ will be a value greater than zero when ΔT is greater than zero. From this, it can be known that the grating wavelength will drift to the long wave direction as the temperature increases.

The energy of the reflection spectrum hardly changed in the temperature range before 850 °C. When the ambient temperature exceeds 850 °C, the internal stress of the fiber will be unevenly distributed, causing changes in the effective period and in the effective refractive index of the grating. Finally, the original refractive index modulation of the fiber grating is changed. Refractive index modulation recombines and weakens resulting in a reduction in reflected spectral energy. The grating reflectance spectra showed very high extinction ratios even without prolonged annealing during the heating process. At present, the reported grating sensors that can measure high temperature need a long regeneration process [21]. The resonant wavelength is linearly fitted, as shown in Figure 8. The sensitivity of the sensor was as high as 14.1 pm/°C in the range from room temperature to 1000 °C, and the linearity could reach 0.994. The wavelength resolution of the demodulator was 0.1 pm. The temperature resolution of the sensor could be obtained by calculation to be 0.007 °C.

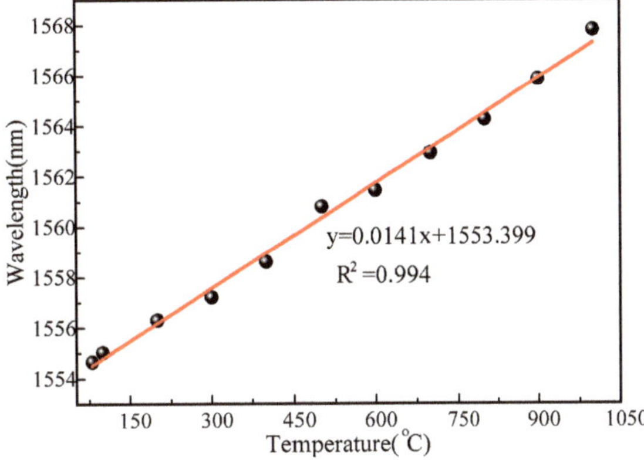

Figure 8. Temperature fitting curve.

The temperature stability of the FBG was continuously monitored for a long time. The output spectrum of the sensor was monitored for 10 h when the temperature was 600 °C and 1000 °C, respectively. The implementation results are shown in Figure 9. The maximum wavelength fluctuation of the reflection spectrum was 18 pm, and the corresponding temperature fluctuation did not exceed 1.2 °C. This fluctuation had little effect compared with the absolute value of the measured temperature.

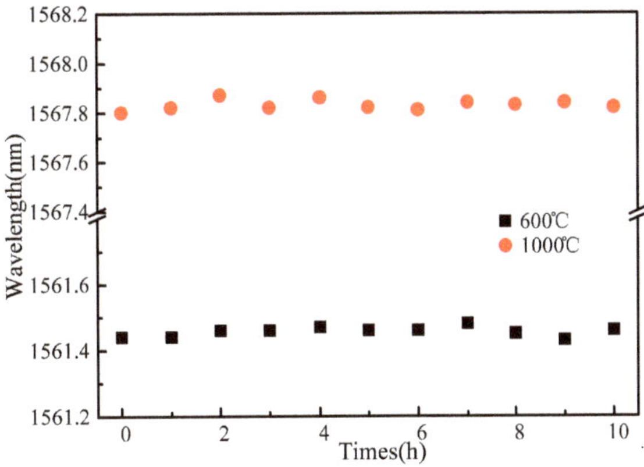

Figure 9. Temperature stability experiment.

There have been reports about the high sensitivity of fiber grating temperature sensors, but there are few reports about its response time characteristics. The response time of the sensor to temperature is an important parameter of the sensor's sensing index. Therefore, a response time test experiment was conducted: instantly switch the sensor from 25 °C to

980 °C high temperature environments; the wavelength signal of the sensor is perceived by the demodulation system, and its sampling frequency is 1 kHz; as shown in Figure 10, the response time of the sensor is 0.6 s for the temperature jump of 955 °C. Since manual switching will cause a specified delay when the sensor switches between two different temperature environments, the actual response speed of the temperature sensor should be faster. The influence of sensor measurement error caused by inconsistent sensor response is eliminated. The response of the traditional platinum resistance temperature sensor is more than 10 s. Although the response time of the thermistor can be very short, the current is difficult to control during the test, and it is often very large, which will cause substantial measurement errors. The extra fiber sensor shows an extremely fast measurement response.

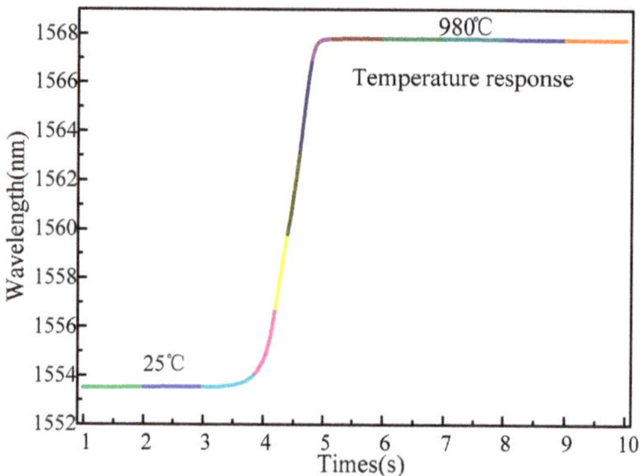

Figure 10. Temperature response time.

Table 3 summarizes the comparison of sensing characteristics of different types of sensors. It can be found from the table that the sensor designed in this paper exhibits excellent sensing characteristics.

The resolution comparison between the sensor in the paper and the same type of fiber optic sensor is given in Table 4. It is obvious from the table that the sensor designed in this paper exhibits a very high measurement resolution.

Table 3. Comparison of sensing characteristics of different types of sensors.

Sensors Structure	Bending Sensitivity	Range	Temperature Sensitivity	Range	Distinguishing Measurement	Reference
Four-Core Sapphire-Derived Fiber	4.5 dB/m^{-1}	0.4–2.5 m^{-1}	/	/	No	[4]
Seven-core FBG	7.2 dB/m^{-1}	0–1.0 m^{-1}	12 pm/°C	35–215 °C	No	[5]
Optical fiber laser	1.04 nm/m^{-1}	0.8–2.0 m^{-1}	/	/	No	[6]
Two-core MZI	−6.18 nm/m^{-1}	0–0.98 m^{-1}	31 pm/°C	30–70 °C	No	[7]
Three core LPFG	3.23 nm/m^{-1}	0–0.58 m^{-1}	54 pm/°C	30–80 °C	No	[8]
Concave-lens-like LPFG	32.78 nm/m^{-1}	0–2.08 m^{-1}	54 pm/°C	30–90 °C	No	[12]
SMF-FCF-SMF	−18.75 nm/m^{-1}	0.042–0.163 m^{-1}	74 pm/°C	30–80 °C	Yes	[15]
Off-axis written FBG	−1.25 dB/m^{-1}	0–1.1 m^{-1}	10.8 pm/°C	23.5–60 °C	No	[19]
This work	21.85 dB/m^{-1}	0–1.2 m^{-1}	14.1 pm/°C	20–1000 °C	Yes	

Table 4. Resolution comparison between different types of sensors.

Sensor Structure	Bending Resolution	Temperature Resolution	Reference
Four-Core Sapphire-Derived Fiber	$0.008\ \mathrm{m}^{-1}$	\	[4]
Seven-core FBG	$0.001\ \mathrm{m}^{-1}$	$0.08\ ^\circ\mathrm{C}$	[5]
Optical fiber laser	$0.004\ \mathrm{m}^{-1}$	\	[6]
Two-core fiber taper	$0.003\ \mathrm{m}^{-1}$	$0.3\ ^\circ\mathrm{C}$	[7]
Three-core LPFG	$0.006\ \mathrm{m}^{-1}$	$0.35\ ^\circ\mathrm{C}$	[8]
Concave-lens-like LPFG	$0.0006\ \mathrm{m}^{-1}$	$0.32\ ^\circ\mathrm{C}$	[12]
graded index multimode fiber	\	$0.11\ ^\circ\mathrm{C}$	[18]
Off-axis	$0.039\ \mathrm{m}^{-1}$	\	[19]
This work	$0.0004\ \mathrm{m}^{-1}$	$0.007\ ^\circ\mathrm{C}$	

4. Conclusions

In this paper, a new type of optical fiber based on different doping materials is designed first, and the refractive index of the core of the optical fiber exhibits a quasi-linear distribution. The ratio and the effective refractive index of different doping materials for the new optical fiber are analyzed. Fiber gratings were engraved on the new fiber by femtosecond laser. The grating model was analyzed by quantum optics theory and refractive index perturbation theory. Bending and temperature experiments were performed on the sensor. The bending experiment results show that the sensitivity of the sensor is as high as $21.85\ \mathrm{dB/m}^{-1}$ in the curvature range of 0 to $1.2\ \mathrm{m}^{-1}$, and its linearity is 0.994. Stability experiments show that the long-term experimental bending fluctuation measurement error is $0.014\ \mathrm{m}^{-1}$. The temperature experiment results show that the grating sensitivity is $14.1\ \mathrm{pm/^\circ C}$ in the temperature range from $20\ ^\circ\mathrm{C}$ to $1000\ ^\circ\mathrm{C}$, and its linearity is 0.994. At the same time, the sensor shows excellent stability. The sensing system has extremely high bending and temperature measurement resolution. The grating does not require long-term annealing to enable high temperature measurements. The temperature response time of the sensor is 0.6 s for the temperature jump of $955\ ^\circ\mathrm{C}$. Due to excellent sensing properties, it has important potential applications in extreme conditions.

Author Contributions: Conceptualization, Y.Z. (Yunshan Zhang), Y.Z. (Yulin Zhang), X.H. and L.K; methodology, Y.Z. (Yunshan Zhang); validation, Y.Z. (Yunshan Zhang), Y.Z. (Yulin Zhang) and D.W.; formal analysis, D.W.; investigation, L.F.; data curation, Z.W.; writing—original draft preparation, Y.Z. (Yunshan Zhang); writing—review and editing, L.K.; supervision, Y.Z. (Yulin Zhang) and L.K.; project administration, Y.Z. (Yulin Zhang) and L.K.,; funding acquisition, Y.Z. (Yunshan Zhang) and L.K. All authors have read and agreed to the published version of the manuscript.

Funding: This research was funded by the China Postdoctoral Science Foundation (grant number 2021M691742), Equipment Development Department support project (grant number 30409), National Natural Science Foundation of China (grant number 62105278), and Natural Science Foundation of Shandong Province under Grant No. ZR2021QF139.

Institutional Review Board Statement: Not applicable.

Informed Consent Statement: Not applicable.

Data Availability Statement: Not applicable.

Conflicts of Interest: The authors declare no conflict of interest.

References

1. Bao, W.; Sahoo, N.; Sun, Z.; Wang, C.; Liu, S.; Wang, Y.; Zhang, L. Selective fiber Bragg grating inscription in four-core fiber for two-dimension vector bending sensing. *Opt. Express* **2020**, *28*, 26461. [CrossRef] [PubMed]
2. Lei, X.Q.; Feng, Y.; Dong, X.P. High-temperature sensor based on a special thin-diameter fiber. *Opt. Commun.* **2020**, *463*, 125386. [CrossRef]
3. Reyes-Vera, E.; Cordeiro, C.M.B.; Torres, P. Highly sensitive temperature sensor using a Sagnac loop interferometer based on a side-hole photonic crystal fiber filled with metal. *Appl. Opt.* **2017**, *56*, 156–162. [CrossRef] [PubMed]

4. Wang, Z.; Zhang, L.; Ma, Z.; Chen, Z.; Wang, T.; Pang, F. High-Sensitivity Bending Sensor Based on Supermode Interference in Coupled Four-Core Sapphire-Derived Fiber. *J. Light. Technol.* **2021**, *39*, 3932–3940. [CrossRef]
5. Zhang, Y.; Zhang, W.; Zhang, Y.; Wang, S.; Yu, L.; Yan, T. Simultaneous measurement of curvature and temperature based on LP 11 mode Bragg grating in seven-core fiber. *Meas. Sci. Technol.* **2017**, *28*, 055101. [CrossRef]
6. Xiong, H.; Xu, B.; Wang, D.N. Temperature Insensitive Optical Fiber Laser Bend Sensor with a Low Detection Limit. *IEEE Photonics Technol. Lett.* **2015**, *27*, 2599–2602. [CrossRef]
7. Wang, L.; Zhang, Y.; Zhang, W.; Kong, L.; Li, Z.; Chen, G.; Yang, J.; Kang, X.; Yan, T. Two-dimensional microbend sensor based on the 2-core fiber with hump-shaped taper fiber structure. *Opt. Fiber Technol.* **2019**, *52*, 101948. [CrossRef]
8. Wang, S.; Zhang, W.; Chen, L.; Zhang, Y.; Geng, P.; Zhang, Y.; Yan, T.; Yu, L.; Hu, W.; Li, Y. Two-dimensional microbend sensor based on long-period fiber gratings in an isosceles triangle arrangement three-core fiber. *Opt. Lett.* **2017**, *42*, 4938–4941. [CrossRef]
9. Bai, Z.; Gao, S.; Deng, M.; Zhang, Z.; Li, M.; Zhang, F.; Liao, C.; Wang, Y.; Wang, Y. Bidirectional bend sensor employing a microfiber-assisted U-shaped Fabry-Perot cavity. *IEEE Photonics J.* **2017**, *9*, 7103408. [CrossRef]
10. Cao, H.; Su, X. Miniature All-Fiber High Temperature Sensor Based on Michelson Interferometer Formed with a Novel Core-Mismatching Fiber Joint. *IEEE Sens. J.* **2017**, *17*, 3341–3345. [CrossRef]
11. Zhao, J.; Zhao, Y.; Bai, L.; Zhang, Y.-N. Sagnac Interferometer Temperature Sensor Based on Microstructured Optical Fiber Filled with Glycerin. *Sens. Actuat. A Phys.* **2020**, *314*, 112245. [CrossRef]
12. Zhang, Y.-S.; Zhang, W.-G.; Chen, L.; Zhang, Y.-X.; Wang, S.; Yu, L.; Li, Y.-P.; Geng, P.-C.; Yan, T.-Y.; Li, X.-Y.; et al. Concave-lens-like long-period fiber grating bidirectional high-sensitivity bending sensor. *Opt. Lett.* **2017**, *42*, 3892–3895. [CrossRef] [PubMed]
13. Mohammed, N.; Serafy, H. Ultra-sensitive quasi-distributed temperature sensor based on an apodized fiber Bragg grating. *Appl. Opt.* **2018**, *57*, 273–282. [CrossRef] [PubMed]
14. Herrera-Piad, L.A.; Hernández-Romano, I.; May-Arrioja, D.A.; Minkovich, V.P.; Torres-Cisneros, M. Sensitivity Enhancement of Curvature Fiber Sensor Based on Polymer-Coated Capillary Hollow-Core Fiber. *Sensors* **2020**, *20*, 3763. [CrossRef] [PubMed]
15. Xu, S.; Chen, H.; Feng, W. Fiber-optic curvature and temperature sensor based on the lateral-offset spliced SMF-FCF-SMF interference structure. *Opt. Laser Technol.* **2021**, *141*, 107174. [CrossRef]
16. Xia, Y.; Hui, D.; Wei, D.; Wei, X. Weakly-coupled multicore optical fiber taper-based high-temperature sensor. *Sens. Actuators A Phys.* **2018**, *280*, 139–144.
17. Xia, P.; Tan, Y.; Li, T.; Zhou, Z.; Lv, W. A high-temperature resistant photonic crystal fiber sensor with single-side sliding Fabry-Perot cavity for super-large strain measurement. *Sens. Actuators A Phys.* **2021**, *318*, 112492. [CrossRef]
18. Niu, H.; Chen, W.; Liu, Y.; Jin, X.; Li, X.; Peng, F.; Geng, T.; Zhang, S.; Sun, W. Strain, bending, refractive index independent temperature sensor based on a graded index multimode fiber embedded long period fiber grating. *Opt. Express* **2021**, *29*, 22922–22930. [CrossRef]
19. Feng, D.; Qiao, X.; Albert, J. Off-axis ultraviolet-written fiber Bragg gratings for directional bending measurements. *Opt. Lett.* **2016**, *41*, 1201–1204. [CrossRef]
20. Xu, X.; He, J.; Liao, C.; Yang, K.; Guo, K.; Li, C.; Zhang, Y.; Ouyang, Z.; Wang, Y. Sapphire fiber Bragg gratings inscribed with a femtosecond laser line-by-line scanning technique. *Opt. Lett.* **2018**, *43*, 4562–4565. [CrossRef]
21. Liu, H.; Yang, H.Z.; Qiao, X.; Hu, M.; Feng, Z.; Wang, R.; Rong, Q.; Gunawardena, D.S.; Lim, K.-S.; Ahmad, H. Strain measurement at high temperature environment based on Fabry-Perot interferometer cascaded fiber regeneration grating. *Sens. Actuators A Phys.* **2016**, *248*, 199–205. [CrossRef]
22. Wang, L.; Guo, Z.; Chi, J.; Wang, Y.; Chen, D. Progress in multipurpose research and development of multiform alumina. *Inorg. Chem. Ind.* **2015**, *47*, 11–15.
23. Mudavakkat, V.; Atuchin, V.; Kruchinin, V.; Kayani, A.; Ramana, C. Structure, morphology and optical properties of nanocrystalline yttrium oxide (Y_2O_3) thin films. *Opt. Mater.* **2012**, *34*, 893–900. [CrossRef]
24. Xie, W.; Li, Y.; Wang, J. Recent Study on Properties and Applications of Y_2O_3 Transparent Ceramics. *Guangzhou Chem. Ind.* **2018**, *46*, 7–9.

 *sensors*

Article

Diagnosis of Bone Mineral Density Based on Backscattering Resonance Phenomenon Using Coregistered Functional Laser Photoacoustic and Ultrasonic Probes

Lifeng Yang [1,2,*], Chulin Chen [1], Zhaojiang Zhang [1] and Xin Wei [1]

1   School of Optoelectronic Science and Engineering, University of Electronic Science and Technology of China, Chengdu 610054, China; charlin@std.uestc.edu.cn (C.C.); z15912188054@gmail.com (Z.Z.); weixin@std.uestc.edu.cn (X.W.)
2   Optoelectronic Imaging and Biophotonics Laboratory, University of Electronic Science and Technology of China, Chengdu 610054, China
*   Correspondence: yanglf@uestc.edu.cn

Abstract: Dual-energy X-ray absorptiometry (DXA) machines based on bone mineral density (BMD) represent the gold standard for osteoporosis diagnosis and assessment of fracture risk, but bone strength and toughness are strongly correlated with bone collagen content (CC). Early detection of osteoporosis combined with BMD and CC will provide improved predictability for avoiding fracture risk. The backscattering resonance (BR) phenomenon is present in both ultrasound (US) and photoacoustic (PA) signal transmissions through bone, and the peak frequencies of BR can be changed with BM and CC. This phenomenon can be explained by the formation of standing waves within the pores. Simulations were then conducted for the same bone μCT images and the resulting resonance frequencies were found to match those predicted using the standing wave hypothesis. Experiments were performed on the same bone sample using an 808 nm wavelength laser as the PA source and 3.5 MHz ultrasonic transducer as the US source. The backscattering resonance effect was observed in the transmitted waves. These results verify our hypothesis that the backscattering resonance phenomenon is present in both US and PA signal transmissions and can be explained using the standing waves model, which will provide a suitable method for the early detection of osteoporosis.

Keywords: backscattering resonance; photoacoustic signal; osteoporosis diagnosis; fracture risk

Citation: Yang, L.; Chen, C.; Zhang, Z.; Wei, X. Diagnosis of Bone Mineral Density Based on Backscattering Resonance Phenomenon Using Coregistered Functional Laser Photoacoustic and Ultrasonic Probes. *Sensors* **2021**, *21*, 8243. https://doi.org/10.3390/s21248243

Academic Editor: Theodore Matikas

Received: 12 October 2021
Accepted: 6 December 2021
Published: 9 December 2021

Publisher's Note: MDPI stays neutral with regard to jurisdictional claims in published maps and institutional affiliations.

1. Introduction

Osteoporosis is a metabolic bone disease that worsens as the individual ages. Affected individuals have a high risk of fractures due to increased bone fragility and increased bone porosity. The World Health Organization (WHO) defines osteoporosis as a bone mineral density (BMD) of 2.5 standard deviations (SDs) or less below the young adult mean. However, most osteoporosis cases are undiagnosed until a fracture occurs, which is costly to treat and is associated with high morbidity. X-ray-based BMD measurement represents the gold standard for osteoporosis diagnosis and fracture risk assessment. Dual-energy X-ray absorptiometry (DXA) is widely used to diagnose osteoporosis [1]. It can expose 60% to 80% of the variability in bone strength and assist in the decision making and monitoring of treatment progression. However, other mechanical factors cannot be detected, such as microstructure, collagen, etc., which are also important in determining the fracture risk of bone. The accuracy of DEXA, especially related to fracture risk, has major limitations [2,3]. Twenty-five percent of people with normal bone density also have a fracture risk, but these people cannot be detected with DXA [4], which leads to doubts regarding the accuracy of DXA. DXA, with bone density as the measurement target, is insensitive to changes in bone collagen and bone microstructure. Simultaneously, there are doubts regarding its potential safety because DXA may destroy the cross-linked structure of bone collagen during the measurement process. In recent years, Wang [5] and Carrin [6] confirmed that bone collagen is an important factor affecting bone structure and elastic modulus. French Sophie [7] attempted to use BMD in combination with trabecular bone score (TBS) to improve DXA's accuracy in osteoporosis detection. Diane et al. [8] found that the combination of TBS and DXA could find 66% to 70% of patients with a fracture risk who were missed by DXA alone. The above cases indicate that it is difficult for DXA to identify patients with early fracture risk, and nearly 30% to 50% of patients are neglected owing to the standard leak point [9]. Some researchers found that the bone density of antlers is very low but that antlers have high toughness [10], and that collagen degradation causes a sharp increase in bone brittleness under the same bone density [11,12]. Willett [13] reported that the cross-linking of collagen is positively correlated with strain toughness, and confirmed that the mechanical properties of bone are related to the characteristics of collagen. New evidence reveals that in addition to BMD, which is a major parameter of bone strength [14,15], other mechanical factors, including micro-architecture, post-yield mechanical properties, and bone collagenous matrix, are also important in determining the fracture risk of bone [9]. The relationship between the organic matrix and the mineral content is important for early fracture prediction, but it cannot be explicitly modeled.

Since osteoporosis is preventable if at-risk populations are detected early, diagnostic techniques are of utmost importance. As osteoporosis is a slowly progressing disease, early detection is critical. Compared with DXA, quantitative ultrasound is a cheap, convenient, and non-radiation-based approach to detecting osteoporosis. Noale [4] investigated 8681 subjects and found that calcaneal Quantitative ultrasound (QUS) was similar to DXA in predicting the risk of fracture caused by osteoporosis. Qin [16] used QUS to detect ultrasound speed and bone machinery and found the correlation between Young's modulus and bone density, respectively. Janne [17] analyzed the direct relationship between several variables of ultrasound signals in the frequency domain and pointed out that bone intensity and ultrasound parameters has a strong correlation at 3.5 MHz. At present, the main reason why QUS cannot be used to define fracture risk levels is that QUS results require new WHO standards. However, due to the damage of DXA to bone, QUS is often used in the investigation of low-risk osteoporosis and evaluation of fracture treatment results [18]. Traditional techniques measure the BMD as an index of bone porosity, whereas more current technologies analyze the bone's microstructure [19,20].

Recently, combined backscatter ultrasound (US) and backpropagation (BR) photoacoustic (PA) measurements have enabled advances in assessing bone integrity [21]. Lashkari [22] reported that a combined photoacoustic and ultrasonic dual mode can be used in the early stage. In comparison to PA, US was capable of generating detectable signals from deeper bone sublayers. However, while US signal variations with changes in the cortical layer were insignificant, PA proved to be sensitive even to minor variations of the cortical bone density [23,24]. A problem that needs to be solved from PA/US dual mode in vitro to in vivo testing is how to remove the influence of the skin and soft tissue on the bone surface on the signal. The key goal of this study was to verify that the BR phenomenon is present in both US and PA signal transmissions through bone, and the peak frequencies of BR can change with BMD and CC, which has the potential to be used to determine which part is from the bone tissue instead of the collagen of the skin in early detection.

2. Materials and Methods

2.1. Experimental Setup

The experiment was set up is shown in Figure 1. It allowed for US and PA tests to be conducted on the same sampling point on the bone. The PA excitation was generated by a CW 808 nm diode laser (Jenopotik AG, Jena, Germany), and the laser intensity was modulated by a software function. The laser fluence on the target surface was about 40 mJ/cm^2 for 5 ms per measurement, which is far below the ANSI safety standard of 1425 mJ/cm^2 for 808 nm laser exposure of the skin. A collimator was used to ensure parallel rays that would spread minimally when propagating toward the sample. The collimated laser beam was 2 mm in diameter on the sample. For generating US waves, a 3.5 MHz focused ultrasonic transducer was used (V382, Olympus NDT Inc., Waltham, MA, USA). The backscattered waves were then detected by a 2.25 MHz focused transducer (V305, Olympus NDT Inc., Waltham, MA, USA). Both transducers and the bone sample were submerged under water for acoustic coupling. The laser beam was set perpendicular to the sample, and its point of incidence was used to adjust the focal point of both transducers to the same point on the sample. The angle between the laser beam and each transducer's center axis was 27° according to a method previously used [22].

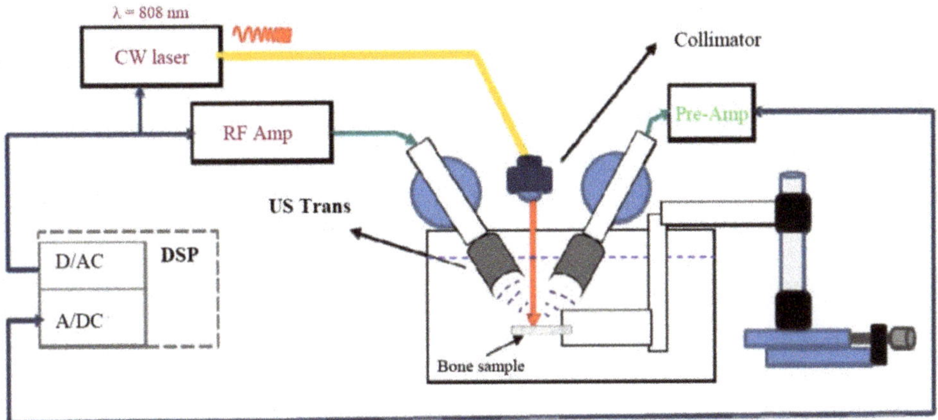

Figure 1. Schematic of the experimental setup.

2.2. Bone Specimens

Bone samples were harvested from the femur and ischium of the same cattle and cut into six rectangular-like samples of similar size. The samples were stored in a refrigerator before processing or measurement, and thermally equilibrated at room temperature before the experiment or measurement. These specimens were randomly divided into two groups and treated with different agents to reduce their mineral or collagen contents [25,26]. Landmarks (two mark points on the sample) were artificially created to distinguish the measurement points before treatment, and then the three samples were treated with 50% ethylenediaminetetraacetic acid (EDTA) buffer solution (pH = 7.7) on the same side of the sample to demineralize the bones, simulating osteoporosis, as shown in Figure 2a. The other three samples were treated with a 5% liquid sodium hypochlorite solution to reduce the CC, as shown in Figure 2b. Landmarks indicated the position of the immersion solution, as shown in Figure 2c, and each measuring point was used as the coordinates (relative position) to indicate the specific position on the sample, as shown in Figure 2d. After preparation, these samples were immersed in phosphate-buffered saline, stored in a refrigerator (−20 °C), and thawed before measurements. The signal of each specimen was detected at the same point before and after sample treatments. The treatment procedures of demineralization and decollagenization were described in detail in the literature [25,26].

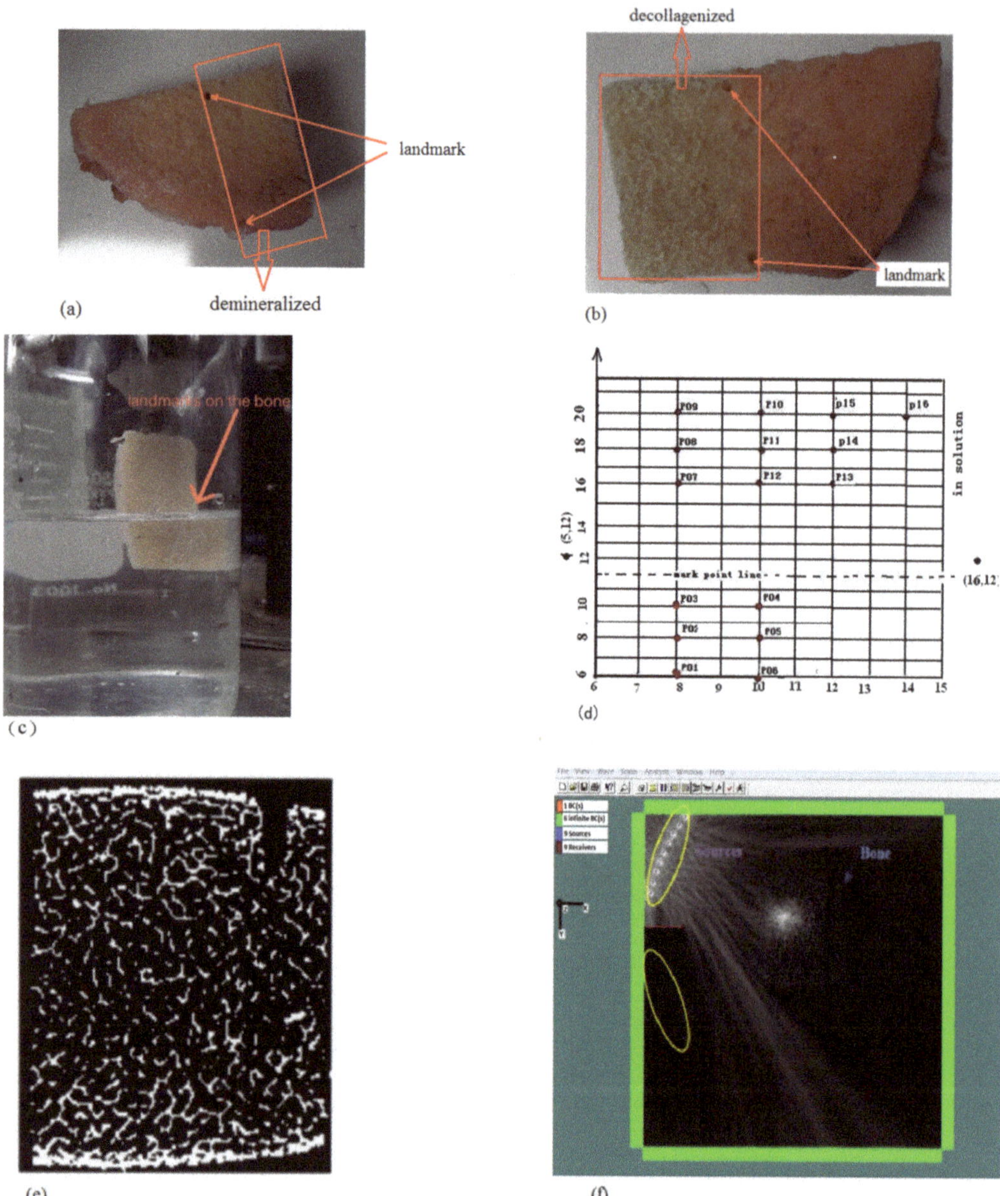

Figure 2. (**a**) Landmarks (two points indicated on the sample) were artificially created to distinguish the measurement points and mark the horizontal line below which the sample was immersed in the solution agent. (**b**) Landmarks (two points indicated on the sample) were artificially marked on the decollagenized sample. (**c**) The indicated line coincides with the surface of the solution. (**d**) The relative position of the 16 measured points landmarked on one sample. (**e**) DICOM image of a bone sample scanned using μCT; the extra void spaces were cropped out. (**f**) Simulations were conducted using a commercial standalone software package for computational ultrasonics (Wave3000®).

2.3. μCT Scanning

On the basis of magnetic resonance imaging (MRI), FE-based estimation of skeletal mechanical capacity involves a number of image-processing and calculation steps [27–29]. The bone sample was first analyzed using micro-computed tomography (μCT) scanning where cross-sectional slices of bone were captured as Digital Imaging and Communications in Medicine (DICOM)images. Each slice was 15 μm thick, and the entire sample yielded 300 images.

The inter distance between the trabeculae of a bone sample along the axis of the source wave was measured by analyzing μCT images using MATLAB (R2020a). The images were then used in Wave3000 (CyberLogic, inc., 2020), which is numerical software for simulating acoustic wave propagation through trabecular bone samples in the defined media and boundaries. This helped to simulate the propagation of ultrasound waves in trabecular bone structure, thereby providing a better understanding of the coherent backscattering effect, enabling analysis of the relationship between resonance frequencies and trabecular bone microstructure [30]. Trabecular bone is considered to consist of two materials: (1) trabeculae: cortical bone (density 1.85 g·cm^{-3}, bulk velocity 2900 m·s^{-1}, and shear velocity 1300 m·s^{-1}); (2) inner and outer coupling medium: water (25 °C, density 1.00 g·cm^{-3}, bulk velocity 1497 m·s^{-1}, and shear velocity 3.54 m·s^{-1}). The values of the acoustic properties for cortical bone and water were obtained from the Wave 3000$^{®}$ material library. Simulation geometry was matched with the experimental measurement geometry. The distance between the transducers was 10 cm and the diameter of the transducers was 2.54 cm. The acoustic source was configured to be identical with the actual transducers (center frequency of 2.25 MHz) used in the experimental measurements [31]. The results of the simulation were then verified experimentally. The purpose of the experiment was to show that the resonance phenomenon [32] is present in bone US and PA signals, thereby justifying the simulation efforts. Further simulations in Wave3000$^{®}$ were conducted by artificially enlarging the μCT images to examine the relationship between peaks in the resonance frequency spectrum and the inter-trabecular distance spectrum, as shown in Figure 2e,f.

2.4. Photoacoustic Backscattering

PA refers to the emission of sound waves from a material after absorbing light waves. Studies showed that a coherent backscattering effect similar to that of US in trabeculae bone can be detected as frequencies higher than 1MHz [25]. Due to the complex nature of bone media, many parameters have been defined for the prediction of the backscattering coefficient (BSC) in the trabeculae. The BSC is a measure of the attenuation of signal caused by scattering at angles from 90° to 180°. Coherent backscattering occurs when the waves propagate through a medium with scattering points of size comparable to the wavelength. This creates the effect termed coherent backscattering because it usually creates a sharp peak in the amplitude vs. frequency graph of the reflected wave in the direction of the backscatter. The frequency of the source wave that generates this coherent backscattering is called the resonance frequency. Some phenomena can explain this sharp peak where standing waves are formed between the inter-trabeculae walls when these waves are scattered multiple times within the trabeculae.

For the formation of standing waves, the resonance frequency can be predicted using the equation for finding standing wave harmonic frequencies:

$$f_n = n\left(\frac{v}{2l}\right) \tag{1}$$

Where f_n is the harmonic frequency, n is the harmonic number, v is the speed of sound in a medium, and l is the length between the nodes, which is the inter-trabecular length.

Equation (1) was used to find the expected resonance frequencies, which were then compared to the resonance frequencies of the simulations and experiments.

2.5. Quantitative Ultrasound (QUS) and Photoacoustic (PA) Measurements

An artificial horizontal landmark line in each sample was used to distinguish the measurement points. The points below this landmark of the sample were immersed in the liquid solution. The points above this landmark were not affected by the solution (Figure 2a,b), so were not demineralized or decollagenized. The points above the landmark were used as a reference to reveal the changes in the US/PA signal due to factors before and after demineralization/decollagenization. The frequency spectra of PA and US signals were normalized by those spectra. In this study, we used the apparent integrated backscatter/back-propagating (AIB) [17,23,31] parameter, which is determined by frequency averaging (integrating) the ratio of the power spectrum of the signal (Pb) over the power spectrum of a reference signal (Pr) over the chirp frequency range:

$$\text{US or PA AIB} = \frac{1}{\Delta f} \int_{\Delta f} 10 \log_{10} \left(\frac{P_b(f)}{P_r(f)} \right) df \tag{2}$$

The apparent US or PA integrated backpropagating signal was calculated using Equation (2), where Pb is the power spectrum of the signal and Pr is the power spectrum of a reference US or PA signal. To eliminate the transfer function effect of the transducer and other instruments, the spectra of the US and PA signals were normalized with their respective reference spectra.

3. Results

3.1. BoneSamples and μCT Scanning

The consolidated inter-trabeculae distances of the 25 slices of bone used in the simulations are shown in Figure 3a. The shape of the distribution is similar to that of a skewed Gaussian distribution centered around 0.6 mm, which was the most frequent value. Note that since this is a histogram, 0.6 mm is the bin value and not the actual value of the inter-trabeculae. The actual values would fall in the range 0.56–0.65 mm. Using 0.6 mm as the value for length in Equation (1) and using 1540 m/s as the speed of 3.5 MHz US in water, the theoretical expected resonance frequencies are f1 = 1.28 MHz and f2 = 2.56 MHz. Only the first two harmonics were considered, since the amplitude decreases exponentially as the harmonic number increases, as shown in Figure 3b.

3.2. Simulation Results

The simulation was completed according to the setup with 1× inter-trabeculae distance bone samples. For each run of the simulation, i.e., for each frequency, the maximum amplitude from each of the nine receivers was identified and averaged across the receivers. The averaged maximum was plotted against the frequency of the source US waves, as shown in Figure 3b.

The original bone sample size showed three peak frequencies, located at 1.3, 1.7, and 1.9 MHz. The first peak at 1.3 MHz is close to the expected resonance frequency of 1.28 MHz calculated previously. If the resonance frequency values of 1.7 and 1.9 MHz are used in Equation (1), the inter-trabeculae distances that could have generated these resonances would be 0.45 and 0.4 mm in length, respectively. This showed that a small peak occurred in the bin value of 0.4 mm, as shown in the histogram in Figure 3b. This could explain why 1.7 and 1.9MHz are also resonance frequencies in this simulation.

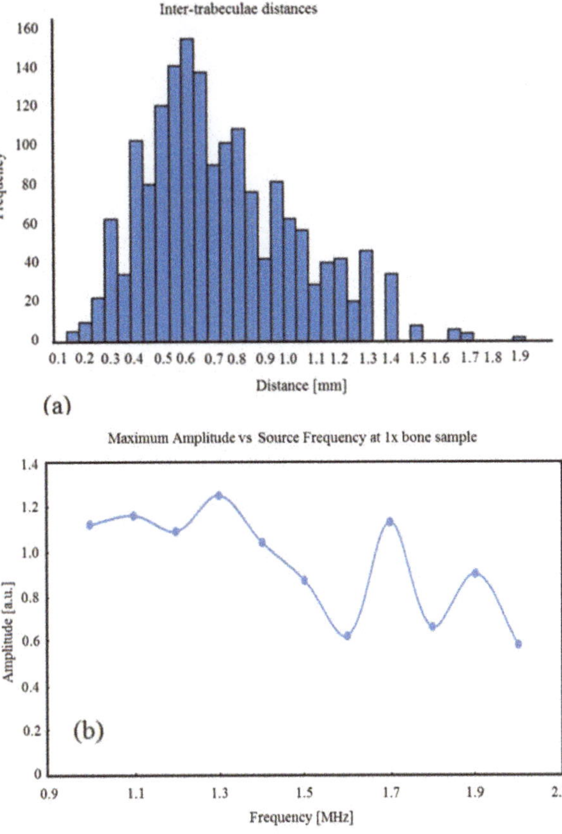

Figure 3. (a) Histogram of inter-trabeculae distances (mm) in the 25 slices of bone used for simulation; (b) maximum amplitude vs. source frequency for the original bone sample.

3.3. Photoacoustic and Ultrasound AIB Comparison

Figure 4 shows that both PA and US AIB values significantly reduced in the demineralized part of the bone. PA signals decreased significantly in the decollagenized part of the bone, whereas US parameters increased slightly in the decollagenized part. Several points of each sample (demineralized or decollagenized) were tested, and the average changes in the US and PA AIB values of each group of samples are shown in Table 1. The averaged correlation coefficients are weak to moderate for the intact parts of all samples with r values ranging from 0.468 to 0.632, which are relatively small compared with those between the US AIB and μCT. Additionally, the PA AIB showed weak correlation with μCT before and after either treatment, as previously reported [22,31].

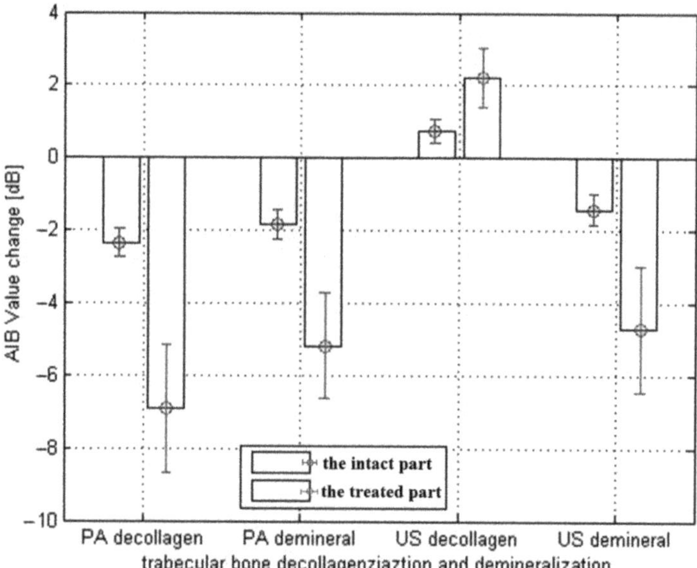

Figure 4. PA and US AIB value variation histogram before and aftertreatment for intact and treated parts of the bone samples.

Table 1. The correlation coefficient between US or PA AIB and microcomputed tomography (μCT) in the treated and intact parts of the samples.

Samples		Intact Part		Treated Part	
		US/μCT	PA/μCT	US/μCT	PA/μCT
Demineralized	1#	0.632	0.327	0.375	0.052
	2#	0.579	0.213	0.252	−0.323
	3#	0.536	−0.107	0.113	−0.357
Average		0.582	0.144	0.247	−0.209
Decollagenized	1#	0.511	0.233	0.394	−0.078
	2#	0.529	0.324	0.287	−0.116
	3#	0.468	0.156	−0.132	−0.096
Average		0.503	0.238	0.183	−0.097

3.4. Actual Experiment Results

Compared with the PA spectrum in the simulation, the PA peak frequencies from the experimental results differed from the peaks in the simulation results both in position and in number. There are only three peaks, at 1.3, 1.7, and 1.9 MHz, in Figure 3b, whereas the PA spectra in Figure 5 have six peaks. This seems to justify the findings of a previous study that showed PA to be more sensitive than US to bone characteristics [31]. Notably, even though the simulated resonance frequency values match the theoretical values used in both simulations, the shape of the frequency spectra differed considerably between the two simulations.

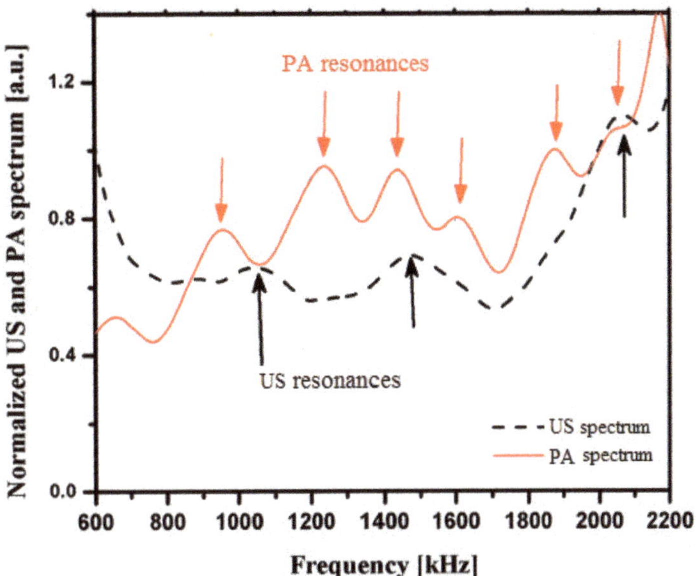

Figure 5. Normalized spectra for US and PA signals vs. source frequency with 1× inter-trabeculae distance bone samples.

Comparing the US spectrum in Figure 5 to that in Figure 3b, a shift in the spectrum can be observed. The three peaks in the simulation results occurred at 1.3, 1.7, and 1.9 MHz, whereas the peaks in the experimental results occurred at 1.1, 1.5, and 2.1 MHz. The differences for all three peaks are ± 0.2 MHz. Arrows in Figures 3b and 5 refer to the positions of the peaks.

By increasing the number of measurement points to 63 on the same sample, we normalized the average of the 63 detected signals to generate the US and PA spectra in Figure 6, where some similar resonances on both PA and US can be observed.

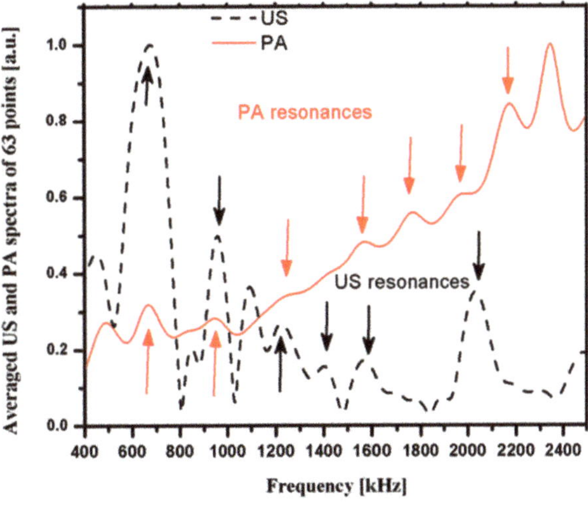

Figure 6. Normalized spectra for US and PA signals vs. source frequency with 63 measurement points.

In Figures 5 and 6, several peaks occur in both the US and PA spectra. This shows that the resonance phenomenon is present in both PA and US signals.

Another experiment was conducted using the treated and intact parts of the same samples, and the results are shown in Figure 7. There are several peaks in both the US and PA spectra in Figure 7a. We found that the PA has a resonance peak similar to that of the ultrasound resonance after demineralization and decollagenization; there is a little difference in the resonance peaks. Conversely, the resonance peaks after demineralization show obvious changes in both the PA and US spectra in Figure 7b, and the resonance peaks after decollagenization have obvious changes in the PA spectra in Figure 7c, indicating that the PA resonance peaks have better sensitivity, especially to changes in collagen. There are important underlying laws that need to be further studied. This seems to justify previous studies that showed PA to be more sensitive to bone characteristics than US.

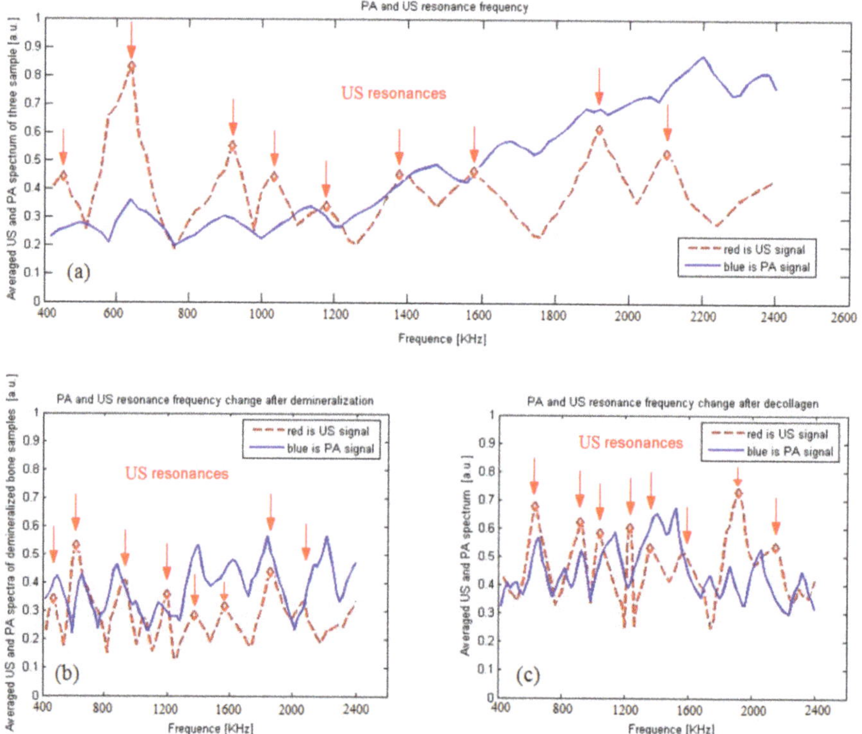

Figure 7. Normalized spectra for US and PA signals vs. source frequency. (**a**) PA and US resonance frequency. (**b**) PA and US resonance frequency change after deminieralization. (**c**) PA and US resonance frequency change after decollagenization.

4. Discussion

Local resonance characteristics occur in PA and US in cancellous bone. The laser is not only absorbed by the collagen in the bone tissue cells, but also by the inorganic calcium bone tissue. The light energy is absorbed by different bone tissues and converted into ultrasonic waves that propagate in the multilayer bone structure.

When sound waves propagate in a composite material with a periodic or quasi-periodic structure, due to the large gap between the acoustic characteristics of two or more materials forming the composite material, the phenomenon of local resonance of the acoustic signal occurs in the deep layer.

This study shows that the PA spectrum exhibits more resonance frequencies or constructive interferences, and there are similar resonance frequencies in PA and US spectra

(Figures 5 and 6). PA is not only sensitive to mineral density variation but also to collagen content. On the other hand, US backscattering is mainly sensitive to BMD variation (Figure 7).

Since simulations are based on solving a single viscoelastic wave equation given some fixed material properties, the aim is to find a solution in an idealized situation where everything happens according to the equation. However, because the trabeculae is such a complex structure, in actual experiments, the US waves may interact in a different manner than dictated by the equation. Variations that could not be accounted for in the simulation, such as inhomogeneous material properties due to varying amounts of mineralization, could be why the experimental results do not match those of the simulation, which needs a better bone mathematical model based on μCT or MRI data to judge the influence of collagen on the signal.

The PA and US dual-mode osteoporosis detection method not only detects the changes in bone density, but also the characteristics of osteoporosis such as the loss of collagen and microstructure trauma, which will provide a suitable method for detecting the early stages of osteoporosis. However, a large amount of collagen exists on the surface of the skin. A problem that needs to be solved from PA/US dual mode in vitro to in vivo testing is how to remove the influence of the skin soft tissue on the bone surface on the signal. If PA and US signals are to detect the information from the collagen in the bone tissue, it is necessary to exclude the influence of the collagen on the skin surface. Local resonance is formed in cancellous bone tissue, but this phenomenon does not occur in surface skin, fat, and other tissues. This article reports, for the first time, the BR phenomenon in both US and PA signal transmissions through the bone, which will provide a more sensitive method for real-time bone diagnostic techniques in vivo.

5. Conclusions

The results from the preceding simulations showed a close correlation between the US resonance frequencies and the first harmonic frequencies of standing waves formed in the most frequently occurring inter-trabeculae distances. The results from the experiments showed the presence of the BR phenomenon in both US and PA signal transmissions through the bone. These results validated our hypothesis that the resonance effect can be observed in both US and PA signal transmissions through the bone, which could be explained by the formation of standing waves. A comparison of the simulation results to experimental results however showed a ±0.2 MHz difference in the resonance frequencies. Therefore, the simulation was not an accurate predictor of the actual experiment. We were able to correctly predict the number of resonance peaks present but the positions of these peaks were inaccurate. Further studies should be conducted in the future to identify more accurate simulation models and to ensure that the resonance effect is dependent only on bone microstructure.

These results verify our hypothesis that the resonance effect can detect the characteristics of osteoporosis, such as the loss of collagen and microstructure trauma, which will provide a suitable method for the early detection of osteoporosis.

Author Contributions: Conceptualization, L.Y. and C.C.; methodology, X.W.; software, X.W.; investigation, X.W.; resources, Z.Z.; data curation, C.C.; writing—original draft preparation, Z.Z.; writing—review and editing, X.W.; supervision, L.Y.; funding acquisition, L.Y. All authors have read and agreed to the published version of the manuscript.

Funding: This work was financially supported by the National Natural Science Foundation of China, grant number 61775030.

Institutional Review Board Statement: Not applicable.

Informed Consent Statement: Not applicable.

Data Availability Statement: The data that support the plots within this paper are available from the corresponding author on request basis.

Acknowledgments: All authors thank Mandelis and Lashkari in CADIFT of Mechanical and Industrial Engineering, of Electrical and Computer Engineering, and of the Institute of Biomedical Engineering, University of Toronto, for helpful discussions.

Conflicts of Interest: The authors declare no conflict of interest.

References

1. NIH Consensus Development Panel on Osteoporosis Prevention, Diagnosis, and Therapy. Osteoporosis prevention, diagnosis, and therapy. *JAMA* **2001**, *285*, 785–795. [CrossRef] [PubMed]
2. Choksi, P.; Jepsen, K.J.; Clines, G.A. The challenges of diagnosing osteoporosis and the limitations of currently available tools. *Clin. Diabetes Endocrinol.* **2018**, *4*, 12. [CrossRef] [PubMed]
3. Geusens, P.; Geel, T.V.; Huntjens, K.; van Helden, S.; Bours, S.; Bergh, J.V.D. Clinical fractures beyond low BMD. *Int. J. Clin. Rheumatol.* **2011**, *6*, 411–421. [CrossRef]
4. Noale, M.; Maqqi, S.; Gonnelli, S.; Limonqi, F.; Zanoni, S.; Rozzini, R.; Crepaldi, G. Quantitative ultrasound criteria for risk stratification in clinical practice: A comparative assessment. *Ultrasound Med. Biol.* **2012**, *38*, 1138–1144. [CrossRef]
5. Wang, X.D.; Ruud, A.B.; Johan, M.K.; Agrawal, C.M. The role of collagen in determining bone mechanical properties. *J. Ortho. Res.* **2001**, *19*, 1021–1026. [CrossRef]
6. Carrin, S.V.; Garnero, P.; Delmas, P.D. The role of collagen in bone strength. *Osteoporos. Int.* **2006**, *17*, 319–336. [CrossRef]
7. Bréban, S.; Briot, K.; Kolta, S.; Paternotte, S.; Ghazi, M.; Fechtenbaum, J.; Roux, C. Identification of rheumatoid arthritis patients with vertebral fractures using bone mineral density and trabecular bone score. *J. Clin. Densitom.* **2012**, *15*, 260–266. [CrossRef] [PubMed]
8. Krueger, D.; Fidler, E.; Libber, J.; Aubry-Rozier, B.; Hans, D.; Binkley, N. Spine Trabecular Bone Score Subsequent to Bone Mineral Density Improves Fracture Discrimination in Women. *J. Clin. Densitom.* **2013**, *15*, 482–487. [CrossRef]
9. Chaffai, S.; Peyrin, F.; Nuzzo, S.; Porcher, R.; Berger, G.; Laugier, P. Ultrasonic characterization of human cancellous bone using transmission and backscatter measurements: Relationships to density and microstructure. *Bone* **2002**, *30*, 229–237. [CrossRef]
10. Charles, H.T. Bone Strength: Current Concepts. *Ann. N. Y. Acad. Sci.* **2006**, *1068*, 429–446.
11. Paschalis, E.P.; Shane, E.; Lyritis, G.; Skarantavos, G.; Mendelsohn, R.; Boskey, A.L. Bone fragility and collagen cross-links. *J. Bone Miner. Res.* **2004**, *19*, 2000–2004. [CrossRef]
12. Wynnyckyj, C.; Willett, T.L.; Omelon, S.; Wang, J.; Wang, Z.; Grynpas, M.D. Changes in bone fatigue resistance due to collagen degradation. *J. Orthop. Res.* **2011**, *29*, 197–203. [CrossRef] [PubMed]
13. Willett, T.L.; Sutty, S.; Gaspar, A.; Avery, N.; Grynpas, M. In vitro non-enzymatic ribation reduces post-yield strain accommodation in cortical bone. *Bone* **2013**, *52*, 611–622. [CrossRef] [PubMed]
14. Launey, M.E.; Buehler, M.J.; Ritchie, R.O. On the mechanistic origins of toughness in bone. *Annu. Rev. Mater. Res.* **2010**, *40*, 25–53. [CrossRef]
15. Georg, E.F.H.; Johannes, H.K.; Hassenkama, T.; James, C.W. Influence of the degradation of the organic matrix on the microscopic fracture behavior of trabecular bone. *Bone* **2004**, *35*, 1013–1022.
16. Qin, Y.X.; Lin, W.; Mittra, E.; Xia, Y.; Cheng, J.; Judex, S.; Rubin, C.; Muller, R. Prediction of trabecular bone qualitative properties using scanning quantitative ultrasound. *Acta Astronaut.* **2013**, *92*, 79–88. [CrossRef]
17. Karjalainen, J.P.; Toyras, J.; Riekkinen, O.; Hakulinen, M.; Jurvelin, J.S. Ultrasound backscatter imaging provides frequency-dependent information on structure, composition and mechanical properties of human trabecular bone. *Ultrasound Med. Biol.* **2009**, *35*, 1376–1384. [CrossRef]
18. Rauma, P.H.; Pasco, J.A.; Berk, M.; Stuart, A.L.; Honkanen, H.K.; Honkanen, R.J.; Hodge, J.M.; Williams, L.J. The association between use of antidepressants and bone quality using quantitative heel ultrasound. *Aust. N. Z. J. Psychiatry* **2015**, *49*, 437–443. [CrossRef]
19. Hakulinen, M.A.; Day, J.S.; Toyras, J.; Timonen, M.; Kroger, H.; Weinans, H.; Kiviranta, I.; Jurvelin, J.S. Prediction of density and mechanical properties of human trabecular bone in vitro by using ultrasound transmission and backscattering measurements at 0.2–6.7 MHz frequency range. *Phys. Med. Biol.* **2005**, *50*, 1629–1642. [CrossRef]
20. Hakulinen, M.A.; Day, J.S.; Toyras, J.; Weinans, H.; Jurvelin, J.S. Ultrasonic characterization of human trabecular bone microstructure. *Phys. Med. Biol.* **2006**, *51*, 1633–1648. [CrossRef]
21. Lashkari, B.; Mandelis, A. Coregistered photoacoustic and ultrasonic signatures of early bone density variations. *J. Biomed. Opt.* **2014**, *19*, 36015. [CrossRef] [PubMed]
22. Lashkari, B.; Yang, L.; Mandelis, A. The application of backscattered ultrasound and photoacoustic signals for assessment of bone collagen and mineral contents. *Quant. Imaging Med. Surg.* **2015**, *5*, 46–56.
23. Lashkari, B.; Mandelis, A. Linear frequency modulation photoacoustic radar: Optimal bandwidth for frequency-domain imaging of turbid media. *J. Acoust. Soc. Am.* **2011**, *130*, 1313–1324. [CrossRef]
24. Lee, K., II; Choi, M.J. Frequency-dependent attenuation and backscatter coefficients in bovine trabecular bone from 0.2 to 1.2 MHz. *J. Acoust. Soc. Am.* **2012**, *131*, EL67–EL73.
25. Ehrlich, H.; Koutsoukos, P.G.; Demadis, K.D.; Pokrovsky, O.S. Principles of demineralization Modern strategies for the isolation of organic frameworks, Part II. Decalcification. *Micron* **2009**, *40*, 169–193. [CrossRef]

26. Hoffmeister, B.K.; Whitten, S.A.; Kaste, S.C.; Rho, J.Y. Effect of collagen content and Mineral Content on the High-frequency Ultrasonic Properties of Human Cancellous Bone. *Osteoporos Int.* **2002**, *13*, 26–32. [CrossRef] [PubMed]
27. Rajapakse, C.S.; Magland, J.F.; Wald, M.J.; Liu, X.S.; Zhang, X.H.; Guo, X.E.; Wehrli, F.W. Computational biomechanics of the distal tibia from high resolution MR and micro-CT images. *Bone* **2010**, *47*, 556–563. [CrossRef] [PubMed]
28. Keyak, J.H.; Sigurdsson, S.; Karlsdottir, G.S.; Oskarsdottir, D.; Sigmarsdottir, A.; Kornak, J.; Harris, T.B.; Sigurdsson, G.; Jonsson, B.Y.; Siggeirsdottir, K.; et al. Effect of finite element model loading condition on fracture risk assessment in men and women: The AGES-Reykjavik study. *Bone* **2013**, *57*, 18–29. [CrossRef]
29. Zhang, H.; Wang, Z.; Wang, L.; Li, T.; He, S.; Li, L.; Li, X.; Liu, S.; Li, J.; Li, S.; et al. A dual-mode nanoparticle based on natural biomaterials for photoacoustic and magnetic resonance imaging of bone mesenchymal stem cells in vivo. *RSC Adv.* **2019**, *9*, 35003–35010. [CrossRef]
30. Nishiyama, K.K.; Ito, M.; Harada, A.; Boyd, S.K. Classification of women with and without hip fracture based on quantitative computed tomography and finite element analysis. *Osteoporos. Int.* **2014**, *25*, 619–626. [CrossRef]
31. Yang, L.; Lashkari, B.; Tan, J.W.Y.; Mandelis, A. Photoacoustic and ultrasound imaging of cancellous bone tissue. *J. Biomed. Opt.* **2015**, *20*, 076016. [CrossRef] [PubMed]
32. Vayron, R.; Nguyen, V.-H.; Lecuelle, B.; Lomami, H.; Meningaud, J.-P.; Bosc, R.; Haiat, G. Comparison of Resonance Frequency Analysis and of Quantitative Ultrasound to Assess Dental Implant Osseointegration. *Sensors* **2018**, *18*, 1397. [CrossRef] [PubMed]

Communication

Theoretical and Experimental Studies of Micro-Surface Crack Detections Based on BOTDA

Baolong Yuan [1], Yu Ying [1,*], Maurizio Morgese [2] and Farhad Ansari [2]

1   College of Information and Control Engineering, Shenyang Jianzhu University, Shenyang 110168, China; kanfenglong@sjzu.edu.cn
2   Department of Civil and Materials Engineering, University of Illinois at Chicago, 842 W Taylor St., Chicago, IL 60607, USA; mmorge3@uic.edu (M.M.); fansari@uic.edu (F.A.)
*   Correspondence: yingyu@sjzu.edu.cn

Abstract: Micro-surface crack detection is important for the health monitoring of civil structures. The present literature review shows that micro-surface cracks can be detected by the Brillouin scattering process in optical fibers. However, the existing reports focus on experiment research. The comparison between theory and experiment for Brillouin-scattering-based optical sensors is rarely reported. In this paper, a distributed optical fiber sensor for monitoring micro-surface cracks is presented and demonstrated. In the simulation, by using finite element methods, an assemblage of a three-dimensional beam model for Brillouin optical time domain analysis (BOTDA) was built. The change in Brillouin frequency (distributed strain) as a function of different cracks was numerically investigated. Simulation results indicate that the amplitudes of the Brillouin peak increase from 27 $\mu\varepsilon$ to 140 $\mu\varepsilon$ when the crack opening displacement (COD) is enlarged from 0.002 mm to 0.009 mm. The experiment program was designed to evaluate the cracks in a beam with the length of 15 m. Experimental results indicate that it is possible to detect the COD in the length of 0.002~0.009 mm, which is consistent with the simulation data. The limitations of the proposed sensing method are discussed, and the future research direction is prospected.

Keywords: distributed optical fiber sensor; micro-cracks; Brillouin frequency

Citation: Yuan, B.; Ying, Y.; Morgese, M.; Ansari, F. Theoretical and Experimental Studies of Micro-Surface Crack Detections Based on BOTDA. *Sensors* **2022**, *22*, 3529. https://doi.org/10.3390/s22093529

Academic Editor: Jin Li

Received: 23 March 2022
Accepted: 3 May 2022
Published: 6 May 2022

Publisher's Note: MDPI stays neutral with regard to jurisdictional claims in published maps and institutional affiliations.

1. Introduction

Optical fiber sensors have been widely used in structural health monitoring for judging the growth process of cracks in buildings, bridges and other forms of civil infrastructure [1–6]. Compared with a traditional electrical sensor, the optical fiber sensor has a lot of advantages, such as anti-corrosion resistance, anti-electromagnetic interference, and high sensitivity. Based on these advantages, optical fiber sensors have been attaining a lot of interest from researchers and engineers in the past few years [7–11].

At present, there are three main optical fiber sensors for structural health monitoring: the local optical fiber sensor, the quasi-distributed optical fiber sensor, and the distributed optical fiber sensor [12–16]. In the aspect of the local optical fiber sensor, Chen H presented an extrinsic Fabry–Perot interferometric (EFPI) with dual cavities to measure the pressure (0.1–3 MPa) and temperature (20–350 °C) [17]. Ghildiyal, S designed an FPI pressure sensor with a copper–beryllium alloy (CBA) diaphragm [18]. A sensitivity of more than 1 μm/bar was obtained. In the aspect of the quasi-distributed optical fiber sensor, Liu, M prepared a fiber Bragg grating (FBG) pressure sensor and encapsulated it in polymer. A sensitivity of 51.296 pm/MPa was experimentally demonstrated in the range of 0~ 15.5 MPa [19]. Guo, G proposed a multiplexed FBG to detect the strain at discrete locations in the fabric. The average lower error rate of 5.9% can be achieved [20]. In the aspect of the distributed optical fiber sensor, Scarella, A presented an optical fiber sensor based on stimulated Brillouin scattering to measure the strain in the bridge model, where the strain increased monotonously with the crack length [21]. Oskoui, E detected five locations of cracks by using Brillouin scattering [22]. Generally speaking, local optical fiber sensors have been limited to a short distance. Quasi-distributed sensors rely on the prior knowledge of crack location. Distributed optical fiber sensors have the advantages of low cost, long distance and, independence from the prior knowledge of crack position. Among them, the use of the distributed optical fiber sensor is a common method for detecting strain caused by crack-growth damages. The method is based on Brillouin scattering, which provides the local strain information in each spatial resolution along the optical fiber. The interaction between light and phones results in Brillouin frequency, which is linearly dependent on strain. This strain will increase as the crack grows. By measuring the Brillouin frequency shift in the optical fiber, the strain data can be acquired, and the crack growth can be identified. In recent years, some significant progress has been made with the development of measurement technology of Brillouin frequency. Yang, D studied a plastic optical fiber sensing technology combining a signal processing method to detect cracks [23]. The results show that the remarkable resemblance in terms of cracks can be identified. Cheng, L designed a high-precision fiber macro-bending loss crack sensor [24]. The macro-bend loss is linearly related to the crack length. Song, Q analyzed a deep learning method and used it for micro-crack detection [25]. The result shows that the crack width of nearly 23 μm can be accurately monitored. Bassil, A presented a multilayer optical fiber to monitor the opening crack in concrete structures [26], with which a relative error as low as 2% can be obtained. This research contributes some novel methods to structural health monitoring. However, most reports are only based on experimental analysis. Less reports focus on the comparison of theory and experiment. Even in the existing theory of crack detection, only the strain exponential model is discussed [26], and it cannot accurately describe the relationship between strain and crack growth. The finite element method of optical fiber can improve the accuracy of analysis, but the analyses of crack detection by use of three-dimensional optical fiber models are fewer. In addition, the tolerance of optical fiber is worth studying.

In this paper, a three-dimensional beam finite element analysis model is proposed, and the strain as a function of crack opening displacement (COD) is analyzed. An experiment program with two CODs, including Brillouin optical time domain analysis (BOTDA) for distributed detection of strain and cracks, is established to evaluate the feasibility. The COD detection range is studied by analyzing the non-linearity of optical fiber.

2. Model Analysis

Accordingly, the present study proposed an optical fiber sensor adhered on a crack substrate (steel beam). Figure 1 shows the distribution optical fiber sensor with two cracks. The theoretical model contains two cracks (COD = $2\delta_1$ and COD = $2\delta_2$) along the optical fiber at $Z = Z_1 = 4.4$ m and $Z = Z_2 = 10.6$ m. The fiber core diameter is 9 μm, the fiber cladding is 125 μm, and the fiber coating is 250 μm.

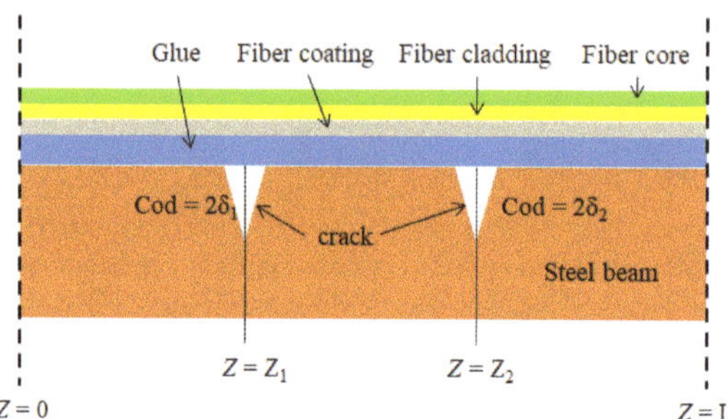

Figure 1. Distribution optical fiber sensor with two cracks.

The system based on Brillouin scattering is achieved by scanning the frequency shift in the optical fiber. External strain applied to the optical fiber can lead to Brillouin frequency shift, and the Brillouin frequency shift $\Omega_B(Z)$ as a function of strain is expressed by [27]

$$\Omega_B(Z) = C\varepsilon\Delta\varepsilon(Z) + C_T\Delta T(Z) \tag{1}$$

where $\Delta\varepsilon(Z)$ and $\Delta T(Z)$ refer to the change in strain and temperature, and $C\varepsilon = 0.05\ \text{MHz}/\mu\varepsilon$ and $C_T = 1\ \text{MHz}/^\circ\text{C}$ are Brillouin factors for strain and temperature, respectively.

The strain with the crack in the distributed optical fiber is expressed by [28]

$$\varepsilon(z) = \begin{cases} \delta_1\beta\exp[\beta(Z-Z_1)] + \delta_2\beta\exp[\beta(Z-Z_2)] & Z < Z_1 < Z_2 \\ \delta_1\beta\exp[-\beta(Z-Z_1)] + \delta_2\beta\exp[\beta(Z-Z_2)] & Z_1 < Z < Z_2 \\ \delta_1\beta\exp[-\beta(Z-Z_1)] + \delta_2\beta\exp[-\beta(Z-Z_2)] & Z_1 < Z_2 < Z \end{cases} \tag{2}$$

where β is the shear lag factor, which was 25/m consulted in the reference [29]. Figure 2 shows the finite element model in which the SMF28 optical fiber was used. The region may have to be divided into 23,546 mesh elements. The steel beam is a rigid body, so it was not used in the model. The cracks were applied to the bottom surface of the glue layer. Since the crack is axisymmetric, the middle section was built in the COMSOL software. The size parameter is $H_g = 6$ mm, $W_g = 10$ mm, and $L_g = 106$ mm. The mechanical properties of the materials can be consulted in the reference [28].

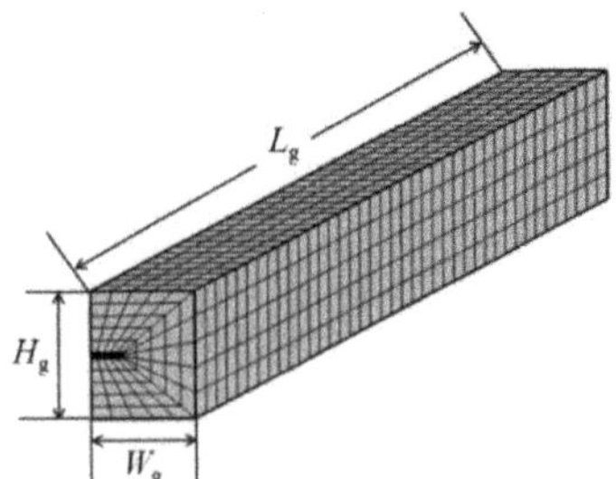

Figure 2. FEM model and mesh dividing of optical fiber coated by glue.

In order to evaluate the feasibility of the presented BOTDA sensor in the long beam with cracks, an experimental program was designed. Figure 3 shows a test bed, which was fabricated to apply the bending test in a 15 m length of beam. The beam was comprised of three sections at the length of 4.4 m, 6.2 m, and 4.4 m, and had two spliced points. The three sections of the beam were connected by bolts and plates. The beam was supported at two points with the span, and the force was applied at the ends. The damage could be fabricated in each splice joint through tightening or loosening the bolts at the plates. In order to monitor the crack opening displacements caused by loosening the bolts, an arch FBG displacement sensor was used. The detailed fabrication method can be consulted in the reference [30]. A single-mode optical fiber (SMF-28) was used as the sensing optical fiber over the beam. In order to maintain stability, the optical fiber was adhered by glue (epoxy resin) onto the top surface of the beam. An available BOTDA measuring system (Neubrex NBX-6055) was used for measuring the strain of the optical fiber. The BOTDA measuring system employs two light sources, including a pump light and a probe light. They transmit in two opposite directions via the separated optical fiber. The sampling interval and spatial resolution were set as SI = 5 cm and SR = 10 cm. The dynamic measurement was conducted at 26 Hz speed (26 measurements/s).

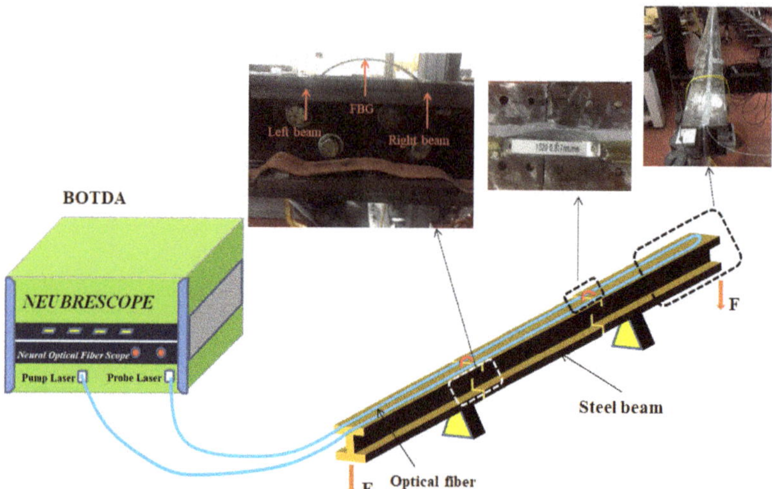

Figure 3. Schematic of experiment setup for detecting cracks with long distributed optical fiber sensors.

Table 1 shows the material parameters of the single-mode optical fiber (single-mode transmission) which is considered a distributed sensor.

Table 1. Mechanical parameters of distributed optical fiber.

Material	Young's Modulus (MPa)	Poisson's Ratio
Fiber core and cladding (silicon dioxide)	72,000	0.2
Fiber coating (acrylate)	4.17	0.48
Glue (epoxy resin)	4000	0.34

3. Results and Discussion

The theoretical and experimental results were studied and compared. For the simulation model, the strain distribution in the optical fiber with two cracks was studied. The distributed optical fiber sensor were assumed to undergo 49 N, 98 N, 196 N, and 392 N, and the crack opening displacements (CODs) were set as 0.002 mm, 0.006 mm, 0.009 mm, and 0.011 mm. Figure 4a shows the Brillouin frequency shift by numerical integration. It can be seen that when the applied force was increased from 49 N to 392 N, the strain in the middle section increased from 10.3 $\mu\varepsilon$ to 60 $\mu\varepsilon$. Two distinct peaks appeared in the Brillouin frequency spectrum due to the two cracks. The amplitudes of the Brillouin peaks increased from 27 $\mu\varepsilon$ to 140 $\mu\varepsilon$ when the CODs at the crack's location rose from 0.002 mm to 0.011 mm. In order to verify the theoretical analysis, distributions of strain were measured for the four CODs ranging from 0.002 mm to 0.011 mm. Figure 4b shows the distributions of strain along the length of the optical fiber from the experiments after filtering out the noise.

It seems that the theoretical results from Figure 4a and the experimental results from Figure 4b are in line with the location of the crack. However, the difference between amplitudes of strain in no-crack and crack regions can be observed when the simulation results are compared with the experimental results. Compared with the actual measurement value, the theoretical simulations underestimate the amplitudes of strain. Figure 5a,b show the difference, respectively. One of the reasons for this difference can be attributed to the crack displacement of the optical fiber. When the crack occurs, the displacement was detected by FBG. However, the actual displacement of the optical fiber was larger than that detected by FBG. Because the optical fiber was not completely attached to the beam surface, there was still a gap above the beam surface. Therefore, the strain was larger than that in the experiment due to the longer crack opening displacement. The other reason is attributed to the fluctuation of the Brillouin scattering gain spectrum. Sometimes, the inconsistencies associated with the probe light and pump light may occur due to the noise level in many measurements.

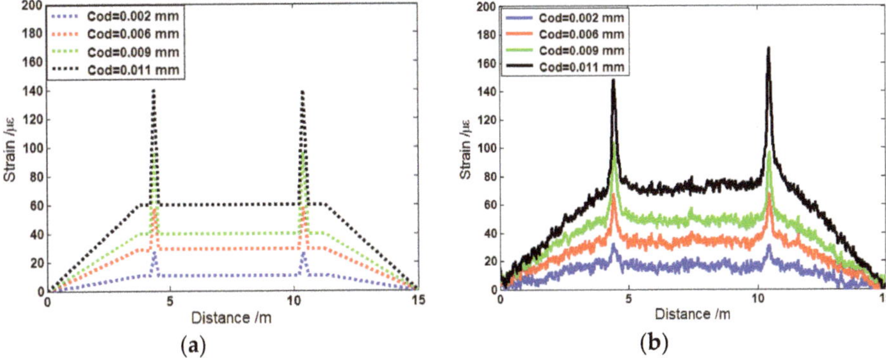

Figure 4. Strain curves with (**a**) theoretical and (**b**) experimental results for COD = 0.002 mm, 0.006 mm, 0.009 mm, and 0.011 mm.

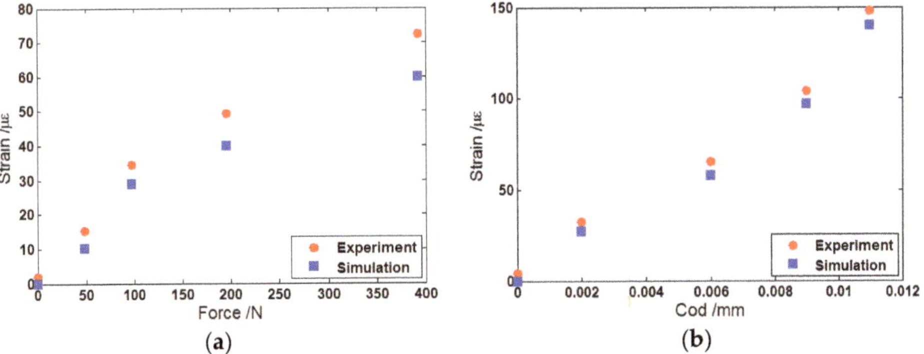

Figure 5. (**a**) Comparison of strain between theoretical and experimental results for *F* = 49 N, 98 N, 196 N, and 392 N. (**b**) Comparison of strain between theoretical and experimental results for COD = 0.002 mm, 0.006 mm, 0.009 mm, and 0.011 mm.

In order to ensure that the optical fiber can be re-used, the experiment was conducted for a second time. Figure 6 shows the comparison of the peaks in the crack strain region for the first time and second time. It was seen that the strain decreases at first and then increases along the crack region. This is attributed to the non-linearity of Young's modulus of the optical fiber [31]. The strain response of optical fiber under continuous loading can be divided into two phases. In the first phase, the optical fiber can exhibit full strain recovery. In the second phase, residual strain occurs in the optical fiber and increases when the strain exceeds a certain threshold value. Figure 7 shows the relation between stress σ and strain ε for distributed optical fiber adhered to the beam surface. When the optical fiber performs in the red region (OA line), the stress and strain exhibits a linear trend. When the strain is larger than the yield stress, the optical fiber performs under a non-linear condition (AB line). In the two ranges of OA and AB, the optical fiber can return to its original memory shape once the stress is released. When the stress exerted upon the optical fiber reaches up to the value of spot B, the optical fiber is physically damaged and cannot return to its original shape, even in the no-stress condition. There will be plastic strain (OC line). In the experiment, due to the excess deformation of the optical fiber, the optical fiber was stretched again, and the stress in the central position decreased. Therefore, the crack below 0.009 mm can be used for the distributed optical fiber through testing the optical fiber characteristic.

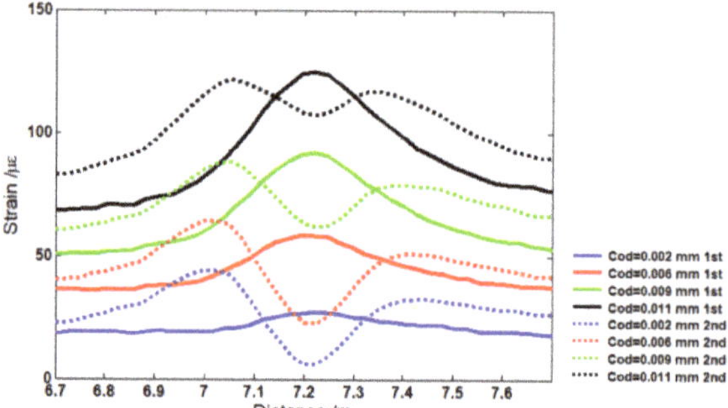

Figure 6. Comparison of first-time and second-time measurement strains in crack region.

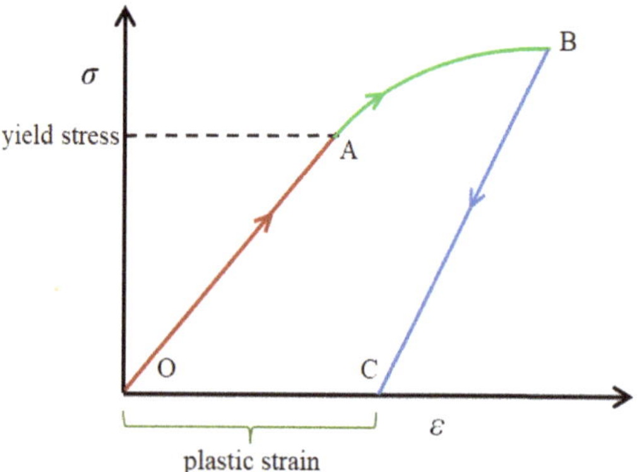

Figure 7. Non-linearity of distributed optical fiber.

4. Conclusions

This paper reports a method for COD determination based on calculating the Brillouin frequency peak. The work involved a theoretical simulation as well as an experimental investigation. In the theoretical perspective, FEM was used for determining the distributed strain along the optical fiber. The simulation results indicate that the Brillouin peak frequency increases from 27 $\mu\varepsilon$ to 140 $\mu\varepsilon$ with an increase in COD from 0.002 mm to 0.009 mm. In the experimental perspective, a 15 m length of steel beam was designed to realize the distributed strain measurement and simulate crack detection. The experimental results are very consistent with the simulation data. In addition, the nonlinear tolerance of the proposed optical fiber was analyzed.

In summary, the proposed Brillouin distributed optical fiber sensor is applicable for detection and monitoring of crack damage in steel beams, with the advantages of long-distance monitoring, low cost, and simple structure. However, this work is in its initial stage, and studies some basic problems in theoretical and experimental aspects. In the future, some research directions require more attention. For example, firstly, optimizing characteristics and practical application. Secondly, the stress–strain non-linearity of the optical fiber core needs to be studied deeply. Thirdly, the influence of the protective coating and glue on strain transferring to the core is a factor worth considering. Fourthly, the range extension of crack detection needs to be further investigated. Finally, it is necessary to consider the error of the optical fiber due to environmental disturbances.

Sensors **2022**, *22*, 3529

Author Contributions: Conceptualization, B.Y.; methodology, software, hardware, validation, and data curation, M.M.; formal analysis, investigation, and resources, Y.Y.; writing—original draft preparation, B.Y.; writing—review and editing, F.A. and M.M.; supervision and project administration, Y.Y. All authors have read and agreed to the published version of the manuscript.

Funding: This research was funded by China Scholarship Council (CSC), grant number (201908210075).

Institutional Review Board Statement: Not applicable.

Informed Consent Statement: Not applicable.

Data Availability Statement: Not applicable.

Acknowledgments: The authors acknowledge the partial financial support from the China Scholarship Council (CSC), project 201908210075.

Conflicts of Interest: The authors declare no conflict of interest.

References

1. Joe, H.E.; Yun, H.; Jo, S.H.; Jun, M.; Min, B.K. A review on optical fiber sensors for environmental monitoring. *Int. J. Precis. Eng. Manuf.-Green Technol.* **2018**, *5*, 173–191. [CrossRef]
2. Li, J.; Yan, H.; Dang, H.; Meng, F. Structure design and application of hollow core microstructured optical fiber gas sensor: A review. *Opt. Laser Technol.* **2021**, *135*, 106658. [CrossRef]
3. Li, J.; Chen, G.; Meng, F. A fiber-optic formic acid gas sensor based on molybdenum disulfide nanosheets and chitosan works at room temperature. *Opt. Laser Technol.* **2022**, *150*, 107975. [CrossRef]
4. Wang, X.; Ansari, F.; Meng, D.; Bao, T. Acousto-opto-mechanical theory for polarization maintaining optical fibers in Brillouin based sensing. *Opt. Fiber Technol.* **2015**, *21*, 170–175. [CrossRef]
5. Motamedi, M.H.; Feng, X.; Zhang, X.; Sun, C.; Ansari, F. Quantitative investigation in distributed sensing of structural defects with Brillouin optical time domain reflectometry. *J. Intell. Mater. Syst. Struct.* **2013**, *24*, 1187–1196. [CrossRef]
6. Meng, D.; Ansari, F. Damped fiber optic low-frequency tiltmeter for real-time monitoring of structural displacements. *Meas. Sci. Technol.* **2013**, *24*, 125106. [CrossRef]
7. Pevec, S.; Donlagic, D. Multiparameter fiber-optic sensors: A review. *Opt. Eng.* **2019**, *58*, 072009.1–072009.26. [CrossRef]
8. Zhao, Y.; Zhao, J.; Zhao, Q. Review of no-core optical fiber sensor and applications. *Sens. Actuators A Phys.* **2020**, *313*, 112160. [CrossRef]
9. Yap, S.H.K.; Chan, K.K.; Zhang, G.; Tjin, S.C.; Yong, K. Carbon dot-functionalized interferometric optical fiber sensor for detection of ferric ions in biological samples. *ACS Appl. Mater. Interfaces* **2019**, *11*, 28546–28553. [CrossRef]
10. Wu, P.; Han, L.; Zeng, N.; Li, F. FMD-Yolo: An efficient face mask detection method for COVID-19 prevention and control in public. *Image Vis. Comput.* **2022**, *117*, 104341. [CrossRef]
11. Zeng, N.; Wu, P.; Wang, Z.; Han, L.; Liu, W.; Liu, X. A small-sized object detection oriented multi-scale feature fusion approach with application to defect detection. *IEEE Trans. Instrum. Meas.* **2022**, *71*, 1–14. [CrossRef]
12. Qi, X.; Wang, S.; Jiang, J.; Liu, K.; Wang, X.; Yang, Y.; Liu, T. Fiber optic Fabry-Perot pressure sensor with embedded MEMS micro-cavity for ultra-high pressure detection. *J. Lightwave Technol.* **2018**, *37*, 2719–2725. [CrossRef]
13. Hong, C.; Yuan, Y.; Yang, Y.; Zhang, Y.; Abro, Z.A. A simple FBG pressure sensor fabricated using fused deposition modelling process. *Sens. Actuators A Phys.* **2019**, *285*, 269–274. [CrossRef]
14. Campanella, C.; Cuccovillo, A.; Campanella, C.; Yurt, A. Fibre Bragg grating based strain sensors: Review of technology and applications. *Sensors* **2018**, *18*, 3115. [CrossRef]
15. Su, H.; Wen, Z.; Li, P. Experimental study on PPP-BOTDA-based monitoring approach of concrete structure crack. *Opt. Fiber Technol.* **2021**, *65*, 102590. [CrossRef]
16. Ye, H.; Liu, J.; Zhou, Y.; Huang, R.; Liu, C. Monitoring of Crack Opening Displacement of Steel Structure by PPP-BOTDA-Distributed Fiber Optical Sensors: Theory and Experiment. *Eng. Fract. Mech.* **2022**, *262*, 108275. [CrossRef]
17. Chen, H.; Chen, Q.; Wang, W.; Zhang, X.; Ma, Z.; Li, Y.; Jing, X.; Yuan, S. Fiber-optic, extrinsic Fabry-Perot interferometric dual-cavity sensor interrogated by a dual-segment, low-coherence Fizeau interferometer for simultaneous measurements of pressure and temperature. *Opt. Express* **2019**, *27*, 38744–38758. [CrossRef]
18. Ghildiyal, S.; Balasubramaniam, R.; John, J. Diamond turned micro machined metal diaphragm based Fabry Perot pressure sensor. *Opt. Laser Technol.* **2020**, *128*, 106243. [CrossRef]
19. Liu, M.; Wu, Y.; Du, C.; Wang, Z. FBG-Based Liquid Pressure Sensor for Distributed Measurement with a Single Channel in Liquid Environment. *IEEE Sens. J.* **2020**, *20*, 9155–9161. [CrossRef]
20. Guo, G.; Hackney, D.; Pankow, M.; Peters, K. Shape reconstruction of woven fabrics using fiber bragg grating strain sensors. *Smart Mater. Struct.* **2019**, *28*, 125081. [CrossRef]
21. Scarella, A.; Salamone, G.; Babanajad, S.K.; De, S.A.; Ansari, F. Dynamic Brillouin Scattering–Based Condition Assessment of Cables in Cable-Stayed Bridges. *J. Bridge Eng.* **2017**, *22*, 04016130. [CrossRef]

22. Oskoui, E.A.; Taylor, T.; Ansari, F. Method and monitoring approach for distributed detection of damage in multi-span continuous bridges. *Eng. Struct.* **2019**, *189*, 385–395. [CrossRef]

23. Yang, D.; Li, D.; Kuang, K.S.C. Fatigue crack monitoring in train track steel structures using plastic optical fiber sensor. *Meas. Sci. Technol.* **2017**, *28*, 105103. [CrossRef]

24. Cheng, L.; Li, Y.; Ma, Y.; Tong, F. The sensing principle of a new type of crack sensor based on linear macro-bending loss of an optical fiber and its experimental investigation. *Sens. Actuators A Phys.* **2018**, *272*, 53–61. [CrossRef]

25. Song, Q.; Zhang, C.; Tang, G.; Ansari, F. Deep learning method for detection of structural microcracks by Brillouin scattering based distributed optical fiber sensors. *Smart Mater. Struct.* **2020**, *29*, 075008. [CrossRef]

26. Bassil, A.; Chapeleau, X.; Leduc, D.; Abraham, O. Concrete Crack Monitoring Using a Novel Strain Transfer Model for Distributed Fiber Optics Sensors. *Sensors* **2020**, *20*, 2220. [CrossRef] [PubMed]

27. Nazarian, E.; Ansari, F.; Zhang, X.; Taylor, T. Detection of Tension Loss in Cables of Cable-Stayed Bridges by Distributed Monitoring of Bridge Deck Strains. *J. Struct. Eng.* **2016**, *142*, 04016018. [CrossRef]

28. Meng, D.; Ansari, F. Interference and differentiation of the neighboring surface microcracks in distributed sensing with PPP-BOTDA. *Appl. Opt.* **2016**, *55*, 9782–9790. [CrossRef]

29. Meng, D.; Farhad, A.; Xin, F. Detection and monitoring of surface micro-cracks by PPP-BOTDA. *Appl. Opt.* **2015**, *54*, 4972–4978. [CrossRef]

30. Bassam, A.; Iranmanesh, A.; Ansari, F. A simple quantitative approach for post earthquake damage assessment of flexure dominant reinforced concrete bridges. *Eng. Struct.* **2011**, *33*, 3218–3225. [CrossRef]

31. Daghash, S.M.; Ozbulut, O.E. Characterization of superelastic shape memory alloy fiber-reinforced polymer composites under tensile cyclic loading. *Mater. Des.* **2016**, *111*, 504–512. [CrossRef]

Communication

Characterization of a Mass-Produced SiPM at Liquid Nitrogen Temperature for CsI Neutrino Coherent Detectors

Fang Liu [1], Xiaoxue Fan [1,2], Xilei Sun [2,*], Bin Liu [1], Junjie Li [2], Yong Deng [2], Huan Jiang [2], Tianze Jiang [1] and Peiguang Yan [3]

1. Beijing Key Laboratory of Passive Safety Technology for Nuclear Energy, School of Nuclear Science and Engineering, North China Electric Power University, Beijing 102206, China; liuf@ncepu.edu.cn (F.L.); Fanxx@ncepu.edu.cn (X.F.); liu_bin@ncepu.edu.cn (B.L.); Jiangtz@ncepu.edu.cn (T.J.)
2. State Key Laboratory of Particle Detection and Electronics, Institute of High Energy Physics, Chinese Academy of Sciences, Beijing 100049, China; lijunjie@ihep.ac.cn (J.L.); dengyong@ihep.ac.cn (Y.D.); enginefly@163.com (H.J.)
3. Shenzhen Key Laboratory of Laser Engineering, College of Physics and Optoelectronic Engineering, Shenzhen University, Shenzhen 518060, China; yanpg@szu.edu.cn
* Correspondence: sunxl@ihep.ac.cn; Tel.: +86-10-6177-1677

Abstract: Silicon Photomultiplier (SiPM) is a sensor that can detect low-light signals lower than the single-photon level. In order to study the properties of neutrinos at a low detection threshold and low radioactivity experimental background, a low-temperature CsI neutrino coherent scattering detector is designed to be read by the SiPM sensor. Less thermal noise of SiPM and more light yield of CsI crystals can be obtained at the working temperature of liquid nitrogen. The breakdown voltage (V_{bd}) and dark count rate (DCR) of SiPM at liquid nitrogen temperature are two key parameters for coherent scattering detection. In this paper, a low-temperature test is conducted on the mass-produced ON Semiconductor J-Series SiPM. We design a cryogenic system for cooling SiPM at liquid nitrogen temperature and the changes of operating voltage and dark noise from room to liquid nitrogen temperature are measured in detail. The results show that SiPM works at the liquid nitrogen temperature, and the dark count rate drops by six orders of magnitude from room temperature (120 kHz/mm^2) to liquid nitrogen temperature (0.1 Hz/mm^2).

Keywords: SiPM; breakdown voltage; dark count rate; liquid nitrogen temperature

Citation: Liu, F.; Fan, X.; Sun, X.; Liu, B.; Li, J.; Deng, Y.; Jiang, H.; Jiang, T.; Yan, P. Characterization of a Mass-Produced SiPM at Liquid Nitrogen Temperature for CsI Neutrino Coherent Detectors. *Sensors* **2022**, 22, 1099. https://doi.org/10.3390/s22031099

Academic Editors: Jin Li and Marco Carminati

Received: 22 November 2021
Accepted: 28 January 2022
Published: 31 January 2022

1. Introduction

Silicon photo-multipliers (SiPM) have been developed rapidly in recent years as an effective alternative for conventional Photo-multiplier Tubes (PMT). The SiPM sensor has many excellent characteristics [1], such as: compact size, easy to develop into detector arrays, works under the low bias voltage (V_{bias}) and a strong ability to resist external magnetic and the electric field [2]. The key parameters, including the working voltage, dark count, quantum efficiency and gain for PMT and SiPM, are shown in Table 1 [3]. In addition, SiPM has high photon detection efficiency (PDE) and performs with an excellent single photon resolving ability. Due to these advantages, SiPM arrays are used for long-range high-speed light detection and the ranging (LiDAR) technique to achieve automotive, machine vision and spacecraft navigation [4,5]. The SiPM was used as read-out unit in the Circular Electron Positron Collider (CEPC) experiment for developing accurate measurements of the Higgs Boson [6], and the SiPM arrays were used to detect dark matter at liquid argon temperature [7]. The photon emission of SiPM from the avalanche pulses that were generated has been investigated in dark conditions [8].

Table 1. Key parameters of PMT and SiPM.

Parameter	PMT	SiPM
Working Voltage	>1000 V	30 V~80 V
Dark Count	4000~800,000	10^5~10^6
Photon Detection Efficiency	20%~25%	25%~70%
Gain	10^5~10^6	>10^6

The coherent elastic neutrino-nucleus scattering (CEvNS) method was first theorized by Freedman in 1974 [9,10] and was the dominated interaction for neutrinos in the energy range below 100 MeV. The COHERENT collaboration firstly detected the phenomenon of CEvNS by using CsI(Na) crystal detector to detect the neutrinos from the spallation neutron source at the Oak Ridge National Laboratory (ORNL) in 2017 [11,12]. The neutrinos produced from different neutrino sources presented different energy spectra, and as such the energy of the corresponding recoil nucleus from coherent scattering and the requirements for detection threshold and background are different [13]. The energy of neutrinos from the reactor source is in a low energy range, usually below 10 MeV, and the energy of the corresponding recoil nucleus is about a few keV. In addition, the count rate decreases exponentially as the detection threshold increases. Therefore, the threshold of the detector needs to be lower than 10 keV or less.

In order to achieve the low threshold detector, we propose a low temperature detector scheme using CsI crystals and SiPMs readout. CsI crystals have the highest light yield among mass produced crystals at low temperatures and can reach around 100 photons/keV [14]. However, the output signals of SiPM can effectively suppress its dark noise by controlling SiPM at low temperature, which can help to lower the detection threshold and obtain a high signal-to-noise ratio. The combination of CsI and SiPM can significantly reduce the detector threshold. Low temperatures can effectively suppress the shortcomings of the high dark count rate (DCR) of SiPMs. In older to reach the requirement of a low threshold detector, DCR should be lower than 0.1 Hz/mm^2 at the liquid nitrogen temperature [15]. Selecting a SiPM that can work normally at the liquid nitrogen temperature is a key step in the detector scheme. It is best to be a mass-produced product for easy procurement. ON Semiconductor's J-Series SiPM is a product widely used in the industry, and we have successfully applied it to the GECAM project [15].

In this study, we design a cryogenic system that can cool the SiPM measurement system to liquid nitrogen temperature and keep the SiPM working environment within a certain temperature range. In this research, we investigate this mass-produced SiPMs at liquid nitrogen temperature to determine whether it can be used in a coherent scattering experiment to detect neutrinos at low threshold.

2. Experiment Setup

A set of self-made automatic temperature feedback control cryogenic systems is designed for cooling SiPM in an ease speed, because it will become brittle when the temperature of SiPM changes too fast, as shown in Figure 1a. It consists of a stainless-steel chamber, liquid nitrogen Dewar, temperature controller (AI-808P), temperature sensor, liquid nitrogen tank, solenoid valve and liquid nitrogen pipeline. We use three temperature sensors in this cryogenic system. One sensor (PT100) is placed at the outside bottom of the chamber to serve as the feedback of the temperature controller, and the other two (Probe C1 and C2, with distance of 5 cm) are put inside the chamber nearby SiPM to monitor the temperature in real time. The temperature controller, which adopts the proportional-integral-derivative (PID) control, controls the amount of liquid nitrogen entering the Dewar through the solenoid valve according to the set cooling rate. The photo of the stainless-steel chamber is shown in Figure 1b, with cable penetration and nitrogen replacement ports on the top. Before cooling down, the inside is replaced with nitrogen and sealed to avoid frost. The stable temperature keeping ability of the cryogenic system is important for the measurement accuracy. We measured the temperature retention ability of this system at the setting temperature 0 °C, and recorded the real-time change of temperature within 8000 s. The result indicates that the temperature is stable at 0 °C with a variation range from −0.3 °C to 0.3 °C. The temperature control and cooling curve of the system is shown in Figure 2 and the black solid curve is the setting temperature, with the dash line and dash-dot line showing the value of two temperature probes at different moments. The temperature changes at Probe C1 and C2 are gentle, about 0.5 degrees per minute, and finally reaching a stable plateau with temperature of C1 is −193 degrees, and C2 is −196 degrees. The temperature of C2 is close to the surface temperature of the SiPM circuit board, which represents the temperature of SiPM. The SiPM is J-60035, size 6×6 mm^2 [16].

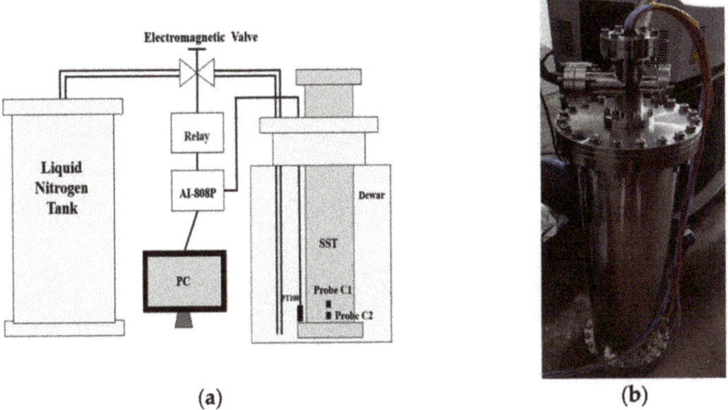

Figure 1. (**a**) Scheme of the cryogenic system; (**b**) photo of the stainless-steel chamber.

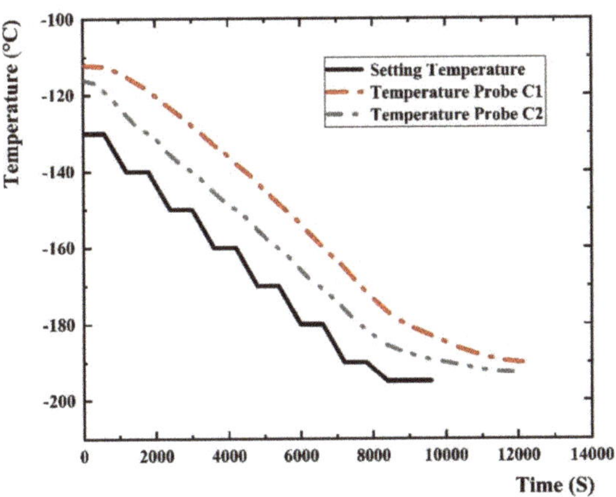

Figure 2. The temperature control and cooling curve of the system.

A Pico-ammeter (Keithley 2450) is used to measure the current versus voltage (I-V) curve of SiPM. In particular, at the liquid nitrogen temperature, an LED (500 nm) light source inside the top of the chamber is turned on; otherwise the breakdown voltage (V_{bd}) of SiPM cannot be measured.

The dark noise measurement scheme is shown in Figure 3. The SiPM signal enters the FIFO (fin in and fin out) (N625) [17] after preamplification, and one fan-out signal enters the low-threshold discriminator (N841) [18] for triggering of the data acquisition system (DT5751) [19], and another fan-out signal is directly connected to the data acquisition system. DT5751 has a counting mode and a waveform acquisition mode, which are used for DCR measurement and single photon spectrum measurement, respectively. The threshold for dark noise measurement is set to 0.5 single photoelectron. The over voltage of SiPM is kept unchanged when comparing noise measurements at different temperatures. The over voltage can be calculated as following:

$$V_{over} = V_{bias} - V_{bd} \tag{1}$$

where V_{over} is the over voltage, V_{bias} is the working voltage and V_{bd} is the breakdown voltage.

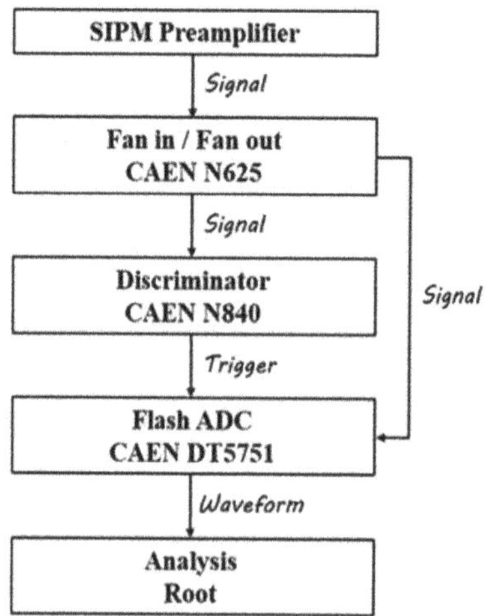

Figure 3. The dark noise measurement scheme.

3. Measurements and Analysis

3.1. I-V Curve

The breakdown voltage (V_{bd}) is the bias point at which the electric field strength in the depletion region is large enough to induce a Geiger discharge. The V_{bd} point is clearly identified on a current versus voltage plot by the sudden increase in current. The I-V curve measured by the Pico-ammeter is shown in Figure 4a, where the typical curves at room temperature are shown. The Pico-ammeter shows the bias of the SiPM, and meanwhile records the currents of SiPM in the working voltage range. We set the step of working voltage change as 0.5 V, and the Pico-ammeter can display the current value varying with the voltage increasing. It can be seen that the V_{bd} of SiPM is around 25 V. Since the step size setting cannot be infinitesimally small, the accurate V_{bd} still needs to be obtained through data processing. The accurate V_{bd} is determined as the value of the voltage intercept of a straight line fit to a plot of $\sqrt{I}$ versus V [20] as shown in Figure 4b, and the fitting result at room temperature is 24.49 V which is very close to the report data (24.7 V) from ON Semiconductor's J-Series SiPM product sheet [17]. It shows that the experimental process and data processing method are reliable and effective. The V_{bd} at different temperatures measured in this way are shown in Figure 5. We can see that the V_{bd} decreases with the decrease of temperature. This change is almost linear when the temperature is higher than $-120\ ^\circ$C, and the change rate is approximately 0.022 V/$^\circ$C. When the temperature becomes lower, the rate of V_{bd} decrease becomes slower.

The thermal vibration of the semiconductor lattice weakens with the decrease of temperature, which results in the widening of the barrier layer in the P-N junction. As such, the mean free path of the carrier movement increases and the energy obtained by the acceleration of the external electric field before colliding with the atom increases [21]. This leads to the enhancement of the chance of collision and ionization, and the probability of avalanche collision increases. Due to this condition, avalanche breakdown is more likely to occur. This is the reason that the V_{bd} becomes lower. Therefore, the V_{bd} of SiPM decreases when the environmental temperature becomes lower. This is consistent with the experimental results.

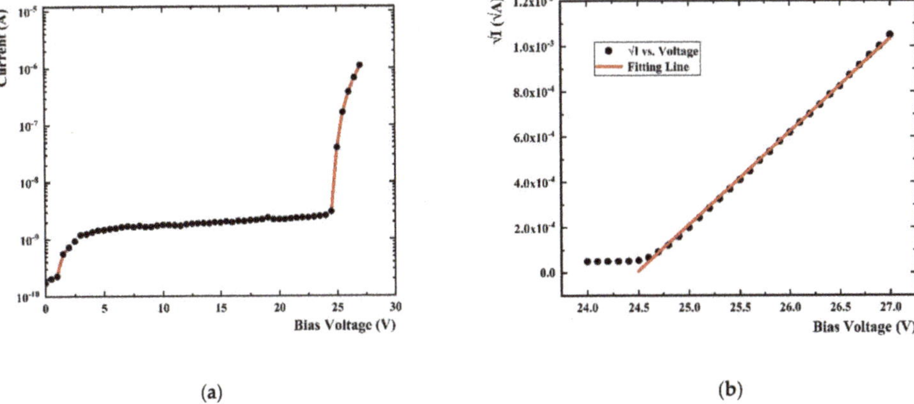

(a)   (b)

Figure 4. (a) Current vs. bias voltage plot at room temperature; (b) $\sqrt{I}/\sqrt{A}$ vs. bias voltage plot at room temperature.

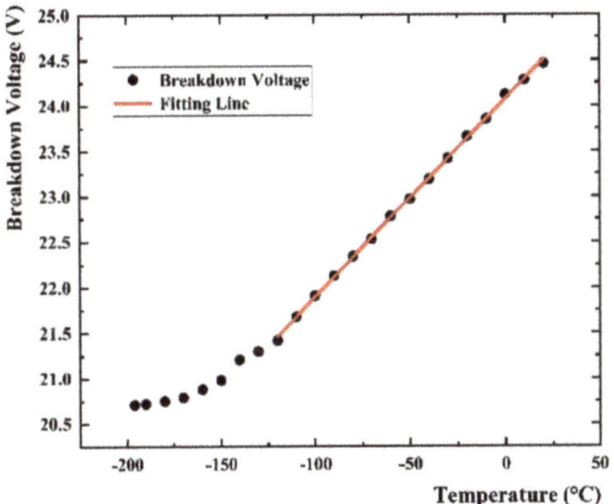

Figure 5. V_{bd} at different temperature, from 20 °C to −196 °C.

3.2. Dark Noise

SiPM has the ability to discern a single photon. Within the scope of certain intensity, the output of SiPM charge is proportional to the photon number and SiPM possess the function of the photon counter. SiPM with better performance can see clear single-photon peaks and multi-photon peaks in the spectrum of dark noise. As shown in Figure 6a, the energy spectra at room temperature and at liquid nitrogen temperature are indicated. In both cases, single-photon peaks can be seen. Since SiPM has lower dark noise at liquid nitrogen temperature, the resolution of single photon peaks is better than that at room temperature. This change shows in the peak-to-valley ratio. The resolution of single photon peak increases with increase of the peak-to-valley ratio. It can be seen from Figure 6 that the peak-to-valley ratio at room temperature is about 6 and this parameter is 12 at liquid nitrogen temperature. The single photoelectron and double photoelectron peaks can be seen clearly in the energy spectrum of SiPM. This indicates that SiPM works perfectly at liquid nitrogen temperature.

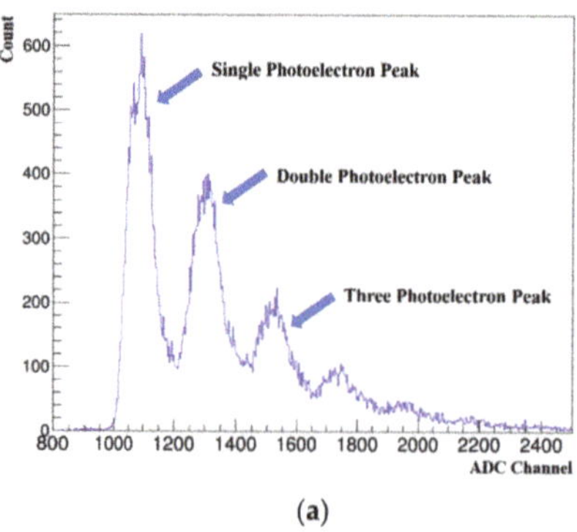

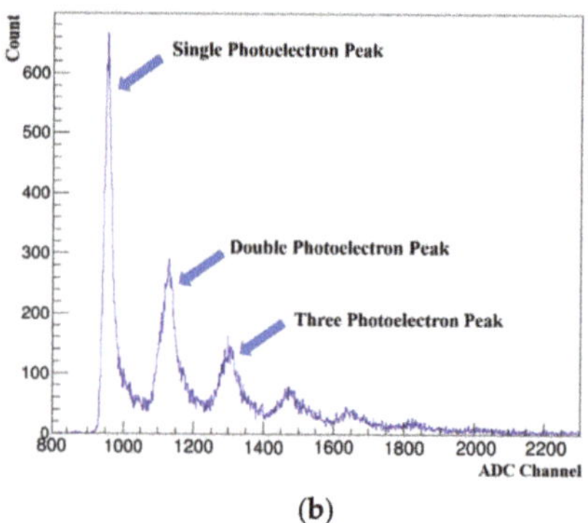

Figure 6. (**a**) Dark noise spectrum of SiPM at room temperature, V_{over} = 2.5 V; (**b**) dark noise spectrum of SiPM at liquid nitrogen temperature, V_{over} = 2.5 V.

Dark noise mainly refers to the dark current. That is, the internal current when the device works in a dark environment. DCR refers to the number of pulses output per unit of time under darkness condition. It directly influences the signal-to-noise ratio of the output signal from SiPM and heavily affects the energy resolution of the crystal scintillator detector. The dark current of SiPM is mainly composed of three parts, which are the surface leakage current, the thermal current caused by the lattice defects in the depletion zone, and the tunneling current caused by the tunnel effect. Dark counting is mainly caused by thermal effects and tunneling effects. It is unavoidable.

Thermal current refers to the current generated by triggering an avalanche during the transition from the valence band to the conduction band of electrons. These electrons are generated by thermal excitation in the depletion zone under the action of an electric field. The calculation formula of the thermal current is as follows:

$$I_{thermal-current} = AT^2 \exp(-E/k_B T) \tag{2}$$

where A is a constant, T is the temperature, E is the energy range between conduction band and valence band, and k_B is Boltzmann constant. It can be seen from the formula that the thermal current is related to the temperature and the current increases with the rise of temperature, and this means a higher DCR. The measured DCR versus temperature curve is shown in Figure 7. DCR dropped by six orders of magnitude from room temperature (120 kHz/mm^2) to liquid nitrogen temperature (0.1 Hz/mm^2). The DCR decreases rapidly with the decrease of temperature from normal temperature to -100 °C, and the decrease gradually becomes slower after -100 °C.

Tunneling current means that the electrons act in a strong electric field to free themselves from the valence band and tunnel to the conduction band, causing an avalanche. The tunneling current is mainly affected by the electric field. The dark current caused by the tunneling effect becomes greater with a higher applied voltage of SiPM, as well as DCR. As shown in Figure 7, when at the same temperature, DCR at V_{over} = 6 V is higher than it at V_{over} = 2.5 V.

The change curve of DCR versus the threshold with 6 V over voltage at liquid nitrogen temperature is shown in Figure 8, and the obvious step structure can be seen, which corresponds to the separation structure of the dark noise spectrum. From the curves shown in Figure 8, the DCR reduces in a step-wise manner with the increase of the threshold. At the stage of low threshold, referring to the threshold of single photoelectron, DCR decreases rapidly. When the threshold range is in the two-photon to three-photon stage, DCR falls into a slower rate. The reason for this phenomenon is that, in the absence of light, most of the collected signals are single photoelectron signals. When the threshold is between single photoelectron and double photoelectrons, the single photoelectron signals get stuck, leaving only a small proportion of multi-photoelectron signals.

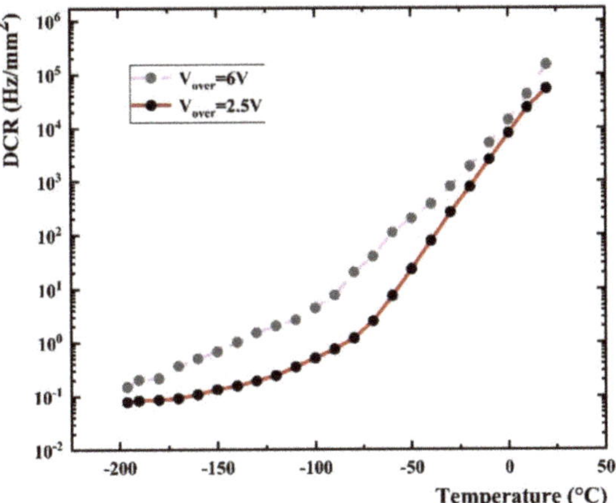

Figure 7. DCR vs. temperature, V_{over} = 2.5 V and V_{over} = 6 V.

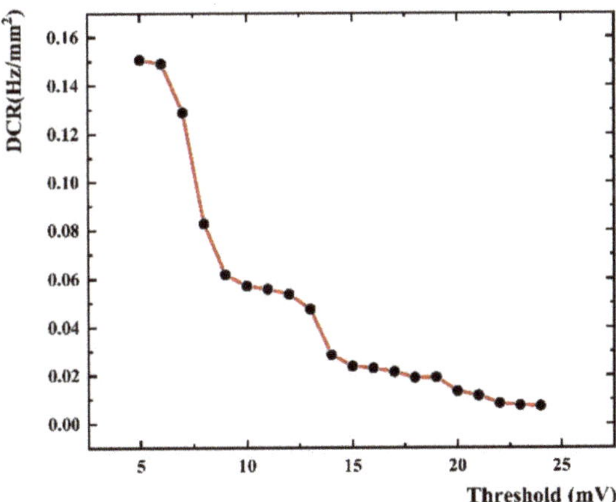

Figure 8. DCR vs. threshold with 6 V over voltage at liquid nitrogen temperature.

4. Conclusions

Knowing the parameters of SiPM, like the V_{bd}, is a prerequisite for using SiPM at different temperatures. The V_{bd} provides a suitable operating voltage range for the use of SiPM. Dark noise will directly affect the signal-to-noise ratio, which is very important in the detection and discrimination of single-photon weak signals using SiPM.

In order to measure the performance of SiPM at liquid nitrogen temperature, we design a cooling system that can maintain the experiment environment at the setting temperature, and achieve a smooth cooling process of SiPM. The experimental results denote that the system can achieve stable control of the setting temperature ranging from the room temperature to liquid nitrogen temperature. The measured result shows that when the setting temperature is 0 °C, the temperature of the control system can fluctuate within the temperature band of −0.3 °C to 0.3 °C. Compared with the conventional temperature, we prefer the ultra-low temperature range. After fitting the experimental result, the rate of decline is less than 0.5 °C/min, which will not cause mechanical damage to the SiPM. We measure the changes of temperature near SiPM when the system temperature decreases, the experimental results show that the mass-produced J series SiPM works properly at liquid nitrogen temperature. The changes of V_{bd} and DCR with temperature are measured in detail; both of them decrease with the decrease of temperature, and the change is relatively fast at the beginning, and gradually slows down after $-100\ ^\circ$C. DCR drops by six orders of magnitude at liquid nitrogen temperature to about $0.1\ Hz/mm^2$ and this meets the requirement of a low threshold detector designing scheme. Preliminary results indicate that the J series SiPMs is one of the candidate devices for neutrino coherent detectors.

Author Contributions: Methodology, X.F. and X.S.; formal analysis, B.L.; investigation, J.L. and X.S.; resources, X.S. and F.L.; data curation, Y.D.; writing—original draft preparation, X.F.; writing—review and editing, X.S. and F.L.; visualization, H.J.; supervision, B.L.; project administration, P.Y. and T.J. All authors have read and agreed to the published version of the manuscript.

Funding: This research was supported by the National Natural Science Foundation of China (11775252) and State Key Laboratory of Particle Detection and Electronics (SKLPDE-ZZ-202116).

Institutional Review Board Statement: Not applicable.

Informed Consent Statement: Not applicable.

Data Availability Statement: Available on request.

Conflicts of Interest: The authors declare no conflict of interest.

References

1. Klanner, R. Characterization of sipms. *Nucl. Instrum. Methods Phys. Res. Sect. A Accel. Spectrometers Detect. Assoc. Equip.* **2019**, *926*, 36–56. [CrossRef]
2. Sun, X.; Tolba, T.; Cao, G.; Lv, P.; Wen, L.; Odian, A.; Vachon, F.; Alamre, A.; Albert, J.; Anton, G.; et al. Study of silicon photomultiplier performance in external electric fields. *J. Instrum.* **2018**, *13*, T09006. [CrossRef]
3. Guo, L.H.; Chen, P.; Li, L.L.; Wu, S.L.; Tian, J.T. Research progress of key technology of photomultiplier tube. *Vac. Electron.* **2020**, 14. [CrossRef]
4. Incoronato, A.; Locatelli, M.; Zappa, F. Statistical Modelling of SPADs for Time-of-Flight LiDAR. *Sensors* **2021**, *21*, 4481. [CrossRef]
5. Villa, F.; Severini, F.; Madonini, F.; Zappa, F. SPADs and SiPMs Arrays for Long-Range High-Speed Light Detection and Ranging (LiDAR). *Sensors* **2021**, *21*, 3839. [CrossRef]
6. Jiang, J.; Zhao, S.; Niu, Y.; Shi, Y.; Liu, Y.; Han, D.; Hu, T.; Yu, B. Study of SiPM for CEPC-AHCAL. *Nucl. Instrum. Methods Phys. Res. Sect. A Accel. Spectrometers Detect. Assoc. Equip.* **2020**, *980*, 164481. [CrossRef]
7. Guo, C.; Guan, M.; Sun, X.; Xiong, W.; Zhang, P.; Yang, C.; Wei, Y.; Gan, Y.; Zhao, Q. The liquid argon detector and measurement of SiPM array at liquid argon temperature. *Nucl. Instrum. Methods Phys. Res. Sect. A Accel. Spectrometers Detect. Assoc. Equip.* **2020**, *980*, 164488. [CrossRef]
8. McLaughlin, J.B.; Gallina, G.; Retière, F.; Croix, A.D.S.; Giampa, P.; Mahtab, M.; Margetak, P.; Martin, L.; Massacret, N.; Monroe, J.; et al. Characterisation of SiPM Photon Emission in the Dark. *Sensors* **2021**, *21*, 5947. [CrossRef] [PubMed]
9. Freedman, D.Z. Coherent effects of a weak neutral current. *Phys. Rev. D* **1974**, *9*, 1389–1392. [CrossRef]
10. Scholberg, K. Observation of coherent elastic neutrino-nucleus scattering by coherent. *J. Phys. Conf. Ser.* **2018**, 1468. [CrossRef]
11. Collar, J.I.; Fields, N.E.; Fuller, E.; Hai, M.; Hossbach, T.W.; Orrell, J.L.; Perumpilly, G.; Scholz, B. Coherent neutrino-nucleus scattering detection with a CsI [Na] scintillator at the SNS spallation source. *Nucl. Instrum. Methods Phys. Res. Sect. A Accel. Spectrometers Detect. Assoc. Equip.* **2015**, *773*, 56–65. [CrossRef]
12. Scholz, B.J. *First Observation of Coherent Elastic Neutrino-Nucleus Scattering*; Recognizing Outstanding Ph.D. Research; Springer: Cham, Switzerland, 2018; Chapter 2; pp. 9–13. [CrossRef]
13. Moszyński, M.; Balcerzyk, M.; Czarnacki, W.; Kapusta, M.; Klamra, W.; Schotanus, P.; Syntfeld, A.; Szawlowski, M.; Kozlov, V. Energy resolution and non-proportionality of the light yield of pure CsI at liquid nitrogen temperatures. *Nucl. Instrum. Methods Phys. Res. Sect. A Accel. Spectrometers Detect. Assoc. Equip.* **2005**, *537*, 357–362. [CrossRef]
14. D'Incecco, M.; Galbiati, C.; Giovanetti, G.K.; Korga, G.; Li, X.; Mandarano, A.; Razeto, A.; Sablone, D.; Savarese, C. Development of a Novel Single-Channel, 24 cm2, SiPM-Based, Cryogenic Photodetector. *IEEE Trans. Nucl. Sci.* **2017**, *65*, 591–596. [CrossRef]
15. Lv, P.; Xiong, S.; Sun, X.; Lv, J.; Li, Y. A low-energy sensitive compact gamma-ray detector based on LaBr3 and SiPM for GECAM. *J. Instrum.* **2018**, *13*, P08014. [CrossRef]
16. J-SERIES SIPM: Silicon Photomultiplier Sensors, J-Series (SiPM). Available online: https://www.onsemi.com/products/sensors/photodetectors-sipm-spad/silicon-photomultipliers-sipm/j-series-sipm (accessed on 21 November 2021).
17. Technical Information Manual of N625. Available online: https://www.caen.it/products/n625/ (accessed on 21 November 2021).
18. Technical Information Manual of N841. Available online: https://www.caen.it/products/n841/ (accessed on 21 November 2021).
19. Technical Information Manual of DT5751. Available online: https://www.caen.it/products/dt5751/ (accessed on 21 November 2021).

20. Introduction to the Silicon Photomultiplier. Available online: https://www.onsemi.com/pub/collateral/and9770-d.pdf (accessed on 21 November 2021).
21. Crowell, C.R. Temperature dependence of avalanche multiplication in semiconductors. *Appl. Phys. Lett.* **1966**, *9*, 242–244. [CrossRef]

91

Article

Study of the Off-Axis Fresnel Zone Plate of a Microscopic Tomographic Aberration

Lin Yang [1,2], Zhenyu Ma [1], Siqi Liu [1,2], Qingbin Jiao [1], Jiahang Zhang [1,2], Wei Zhang [1,2], Jian Pei [1,2], Hui Li [1,2], Yuhang Li [1,2], Yubo Zou [1,2], Yuxing Xu [1,2] and Xin Tan [1,3,*]

[1] Changchun Institute of Optics, Fine Mechanics and Physics, Chinese Academy of Sciences, Beijing 100049, China; yanglinciomp@163.com (L.Y.); emozhiai123@163.com (Z.M.); remiellsq@163.com (S.L.); voynichjqb@163.com (Q.J.); zhangjh0620@163.com (J.Z.); zhangv_ciomp@163.com (W.Z.); peijian980102@163.com (J.P.); lh19990581721@163.com (H.L.); lyhang3111@163.com (Y.L.); z_ouyubo@163.com (Y.Z.); bryant23@126.com (Y.X.)

[2] University of Chinese Academy of Sciences, Beijing 100049, China

[3] Center of Materials Science and Optoelectronics Engineering, Chinese Academy of Sciences, Beijing 100049, China

* Correspondence: xintan_grating@163.com

Citation: Yang, L.; Ma, Z.; Liu, S.; Jiao, Q.; Zhang, J.; Zhang, W.; Pei, J.; Li, H.; Li, Y.; Zou, Y.; et al. Study of the Off-Axis Fresnel Zone Plate of a Microscopic Tomographic Aberration. *Sensors* **2022**, *22*, 1113. https://doi.org/10.3390/s22031113

Academic Editor: Jin Li

Received: 27 December 2021
Accepted: 26 January 2022
Published: 1 February 2022

Publisher's Note: MDPI stays neutral with regard to jurisdictional claims in published maps and institutional affiliations.

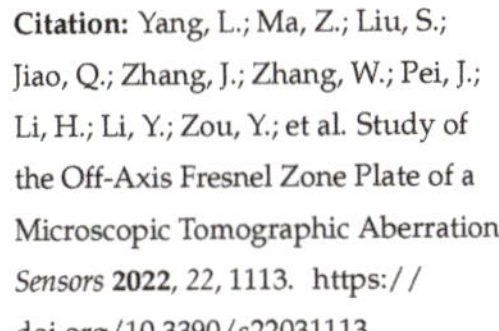

Abstract: A tomographic microscopy system can achieve instantaneous three-dimensional imaging, and this type of microscopy system has been widely used in the study of biological samples; however, existing chromatographic microscopes based on off-axis Fresnel zone plates have degraded image quality due to geometric aberrations such as spherical aberration, coma aberration, and image scattering. This issue hinders the further development of chromatographic microscopy systems. In this paper, we propose a method for the design of an off-axis Fresnel zone plate with the elimination of aberrations based on double exposure point holographic surface interference. The aberration coefficient model of the optical path function was used to solve the optimal recording parameters, and the principle of the aberration elimination tomography microscopic optical path was verified. The simulation and experimental verification were carried out utilizing a Seidel coefficient, average gradient, and signal-to-noise ratio. First, the aberration coefficient model of the optical path function was used to solve the optimal recording parameters. Then, the laminar mi-coroscopy optical system was constructed for the verification of the principle. Finally, the simulation calculation results and the experimental results were verified by comparing the Seidel coefficient, average gradient, and signal-to-noise ratio of the microscopic optical system before and after the aberration elimination. The results show that for the diffractive light at the orders 0 and ±1, the spherical aberration W040 decreases by 62–70%, the coma aberration W131 decreases by 96–98%, the image dispersion W222 decreases by 71–82%, and the field curvature W220 decreases by 96–96%, the average gradient increases by 2.8%, and the signal-to-noise ratio increases by 18%.

Keywords: microtomography; off-axis Fresnel zone plate; eliminating aberration; Seidel coefficient

1. Introduction

The fast and sensitive acquisition of 3D data is one of the main challenges in modern biological microscopy. Most imaging methods, such as laser scanning confocal techniques, achieve optical laminarization by changing the planes of confocalization, which cannot be imaged at the same time because the focal planes are not recorded simultaneously [1]. In 1999, Blanchard et al. first proposed a new type of variable spacing conic diffractive element-off-axis Fresnel zone plate [2] that could be widely used in multiplanar imaging, fluid velocity measurement, particle tracking, and other fields [3]. In 2010, Dalgamo et al. first introduced off-axis Fresnel zone plates into a microscopic imaging system for the multiplanar imaging of cells [4]. The off-axis Fresnel zone plate-based tomographic microscopic imaging system reduced the problems of sample light bleaching and light damage caused by strong incident light, and it was suitable for live-cell imaging because of its real-time image capture, which avoided the problem of error image acquisition at different times caused by excessive cell swimming.

In 2011, Feng Y et al. observed two swimming sperms using a tomographic microscopy system based on off-axis Fresnel zone plates, but there were chromatic aberrations and geometric aberrations in the system that affected the imaging quality of the tomographic microscopy system [5]. To solve the chromatic aberration problem in the system, scholars around the world have carried out a series of studies. Feng Y et al. used a pre-dispersion element to correct chromatic aberration and applied it to a tomographic microscopic imaging system to solve the problem of chromatic aberration affecting imaging quality in 2013 [6]. In 2016, S. Abrahamsson et al. used multiple shining gratings and a multi-faceted prism to compensate for dispersion from the shaft Fresnel [7]. In 2018, Kuan He et al. used an automatic algorithm for three-dimensional reconstruction and then corrected the color difference generated in the system [8]. The analysis of the other aspects of a microscopic system has also achieved significant progress. Yuan Xiangzheng et al. combined polarization multiplexing with the deposition of Fresnel wave tabs, achieving unequal spacing micro-imaging in 2019 [9]. In 2020, Lauren W et al. expanded the imaging field of a tomographic microstructure and the cells that could not enter the field of view by designing the diffraction level [10]. The above research work created a more advantageous microscopic system based on a chromatographic microbial microstructure.

Despite recent progress, there are still aberrations in microscopic tomographic systems that affect their imaging quality. In 2012, S. Abrahamsson et al. reduced the spherical aberration in a system by correcting the out-of-focus phase error in the object plane corresponding to the non-zero diffraction orders; however, other geometric aberrations still exist in this type of system [11]. In 2014, H. Liu et al. extracted the background of the original image obtained using a binary mask and fit it using the least-squares method. The final corrected image was obtained by subtracting this aberration polynomial from the original image [12]. This method corrected the aberration with post image processing. To reduce the influence of aberration in a chromatographic microscopy system, the design of the off-axis Fresnel zone plate, the core element of the chromatographic microscopy system, is proposed in this paper based on the calculation of the aberration coefficient of the optical path function according to the Fermat principle. This design allows for both chromatographic microscopy and systematic aberration reduction. The aberrated off-axis Fresnel zone plates are fabricated with wet etching. After the performance test, the tomographic microscopic optical path is set up for cell image acquisition, and the image indexes such as the signal-to-noise ratio and average gradient are used for quantitative analysis.

2. Design of Fresnel Zone Plate for Aberration Correction

The ability of off-axis Fresnel zone plates to correct for aberrations is achieved by varying the grating inscription distribution. Fresnel zone plates with different inscription densities have different abilities to correct phase aberrations; therefore, in this research, a laminar microscopy system based on an off-axis Fresnel zone plate was first simulated in ZEMAX software to determine the types of aberrations. The results of the ZEMAX simulation show that when the incident light is 530 nm and the chromatography depth is 1.5 μm, there is spherical aberration, coma aberration, astigmatism, field curvature, and distortion in each diffraction order of the chromatography microscope, and the ± 1 diffractive light forms a diffuse spot with a radius of 80 μm on the detector.

2.1. Grid Distribution Design and Simulation Verification

In this research, the optical path difference at any point on the substrate was calculated based on the theory of spherical wave geometry by applying Fermat's principle to the optical range function and performing a series expansion on the function. The final inscribed density function of the aberrated off-axis Fresnel zone plate was obtained [13]. The principle is shown in Figure 1, and the specific mathematical equations are:

$$n_{20} = \frac{\cos^2 \gamma}{r_C} - \frac{\cos^2 \delta}{r_D},$$

(1)

$$n_{30} = \frac{\sin \gamma \cdot \cos^2 \gamma}{r_C^2} - \frac{\sin \delta \cdot \cos^2 \delta}{r_D^2},$$

(2)

$$n_{40} = \frac{4 \sin^2 \gamma \cdot \cos^2 \gamma}{r_C^3} - \frac{4 \sin^2 \delta \cdot \cos^2 \delta}{r_D^3} - \frac{\cos^4 \gamma}{r_C^3} + \frac{\cos^4 \delta}{r_D^3},$$

(3)

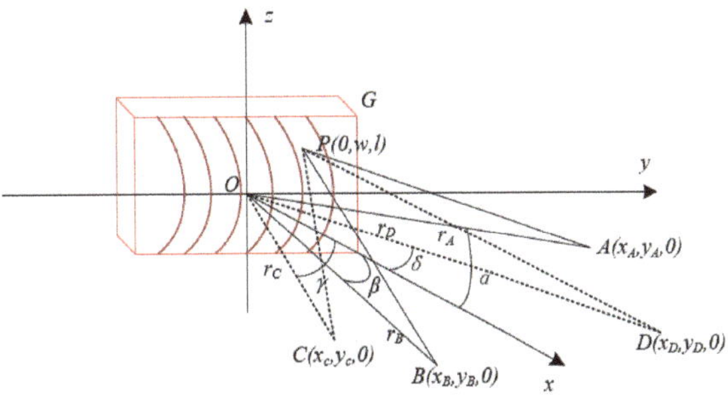

Figure 1. Imaging schematic of aberration axis Fresnel zone plate.

The coefficients n_{20}, n_{30}, and n_{40} are functions of the spherical wave recording parameters (γ, rC, δ, rD), n_{20} is the defocus of the system, n_{30} is the coma of the system, and n_{40} is the spherical aberration. The central period of the aberrated off-axis Fresnel zone plate determines the diffraction angle of each order, which further determines the spacing of each diffraction order [4]. According to the size of the existing CCD (charge-coupled device) and the grating equation, the calculation results show that the center period is $d_0 = 56$ μm and $\delta = 1.47$ rad.

According to the mathematical model of Itou [14], the aberration F_{ij} can be expressed as:

$$F_{20} = \frac{\cos^2 \alpha}{r_A} + \frac{\cos^2 \beta}{r_B} + m \frac{\lambda}{\lambda_0} n_{20},$$

(4)

$$F_{30} = \frac{\sin \alpha \cos^2 \alpha}{r_A} + \frac{\sin \beta \cos^2 \beta}{r_B} + m \frac{\lambda}{\lambda_0} n_{30},$$

(5)

$$F_{40} = \frac{4 \sin^2 \alpha \cos^2 \alpha - \cos^4 \alpha}{r_A{}^3} + \frac{4 \sin^2 \beta \cos^2 \beta - \cos^4 \beta}{r_B{}^3} + m \frac{\lambda}{\lambda_0} n_{40},$$

(6)

Based on the above analysis, the aberration F_{ij} is determined by both the usage parameters and the recording parameters. To use a specific wavelength and the determined usage parameters, the recording parameters can be chosen reasonably so that a specific $F_{ij} = 0$ [15]. Based on the entry arm length of 50 mm, $n_{10} = 18$, $\lambda = 530$ nm, etc., as well as making $F_{20} = 0$, the comet correction equation $F_{30} = 0$ and the spherical aberration correction $F_{40} = 0$, the calculation results show that:

$$n_{10} = 18,$$

(7)

$$n_{20} = 0.214,$$

(8)

$$n_{30} = 0.008327,$$

(9)

$$n_{40} = 0.0002499,$$

(10)

The recording parameters are obtained by substituting in Equations (1)–(3), as shown in Table 1; however, due to the existence of other geometric aberrations in the system, the equations cannot be satisfied simultaneously; therefore, the damped least-squares optimization algorithm is used to optimize the image scattering and field curvature in the region near the calculated results of the recording parameters.

Table 1. Recording parameters of the aberrated off-axis Fresnel zone plate.

	Spherical Aberration, Coma Aberration	Correction System All Aberrations
γ	−1.509 rad	−1.55 rad
δ	1.47 rad	1.47 rad
rC	59.093 mm	133.9396 mm
rD	59.996 mm	133.3251 mm

The premise of the damped least-squares optimization algorithm is to take successive values of the parameters. The value of the evaluation function is made smaller and smaller until the best evaluation function is found. The specific algorithm workflow is:

1. The evaluation function is established. The evaluation function of the laminar microscopy system is the root mean square error of the wavefront. Additionally, in order to avoid the optimization process, the central period is too large, leading to the imaging distance exceeding the CCD size. The central period of 56 μm of the aberrated off-axis Fresnel zone plate is considered to be a nonlinear constraint.
2. The range of values of the angle recording parameter γ is calculated based on the central inscribed density and grating equation of the aberrated off-axis Fresnel zone plate. The optical path displacement range (rCmin, rCmax, rDmin, rDmax) is limited according to the optical stage dimensions being set as the boundary constraints.
3. After inputting the calculated record parameters and starting the operation, the optimal solution of the record parameters is solved by finding the minimum value of the evaluation function at the next point step-by-step around the initial value.

The actual recording parameters are finally obtained, as shown in Table 1. Table 2 shows the parameters of the aberrated off-axis Fresnel zone plate. The number of grid lines n at any point on the aberrated off-axis Fresnel zone plate with the center of the aberrated off-axis Fresnel zone plate as the origin point is shown in Equation (7). Figure 2 represents the overall grid line distribution of the aberrated off-axis Fresnel zone plate. This distribution can be used for the simulation of the laminar microscopy system based on the aberration off-axis Fresnel zone plate in the following, as well as for the subsequent fabrication of the aberration off-axis Fresnel zone plate.

$$n_p = \frac{(DP - CP) - (DO - CO)}{\lambda}, \tag{11}$$

Table 2. The parameters of the aberrated off-axis Fresnel zone plate.

Center Period d_0	Radius R	Thickness b
56 μm	15 mm	2 mm

The simulated optical system of the aberrated off-axis Fresnel zone plate chromatography microscopy system is shown in Figure 3. In the optical microscopy system, the off-axis Fresnel zone plate is replaced by a holographic surface. By inputting the recording parameters calculated in the previous section, the simulation results show that a diffuse spot with a radius of 5 μm is formed at the image plane when the incident light is 530 nm and the lamination depth is 1.5 μm.

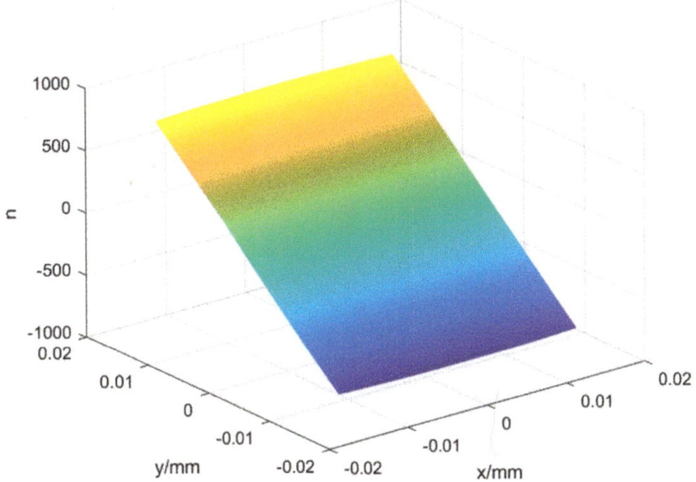

Figure 2. The number of grating lines of the aberrated off-axis Fresnel zone plate.

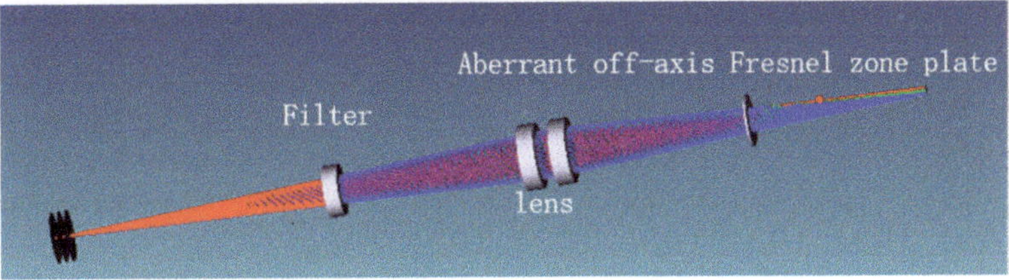

Figure 3. Simulation light path diagram of 3D microscopic imaging system.

From Figure 3, it can be clearly observed that the rays pass through the double-glued lens and grating in turn, and finally, imaging is performed at the image plane; however, the 3D profile does not objectively represent the imaging performance of the system, so it is necessary to analyze the imaging quality of the system with the help of other graphics. The general image quality judgment indexes of optical systems include point column diagrams, light sector diagrams, and Seidel coefficients. The comparison of the aberrations corresponding to different diffraction levels of the unimproved and aberrated optical systems is shown in Figure 4, and the point column diagram corresponding to the two optical systems is shown in Figure 5.

In Figure 4, the two plots in the light sector of each diffraction order represent the aberration in the tangential plane and the sagittal plane. The horizontal coordinates represent the normalized incident pupil, and the vertical coordinate is the value of the ray's deviation from the main ray in the image plane. In the diagram, the larger the vertical coordinate of the peak is, the larger the spherical aberration of the system is. The larger the slope of the aberration curve passing through the origin is, the larger the field curvature in the system is. The longitudinal coordinate of the light sector of the laminar microscope after the system improvement decreases from ± 500 μm to ± 25 μm. This indicates that the spherical aberration is improved due to the counterbalancing of the defocus and spherical aberration. The slope of the aberration curve at the origin is reduced from 0.95 to 0.7, which indicates that the field curvature is also corrected. Furthermore, the position of the image is closer to the ideal image plane with the improved system.

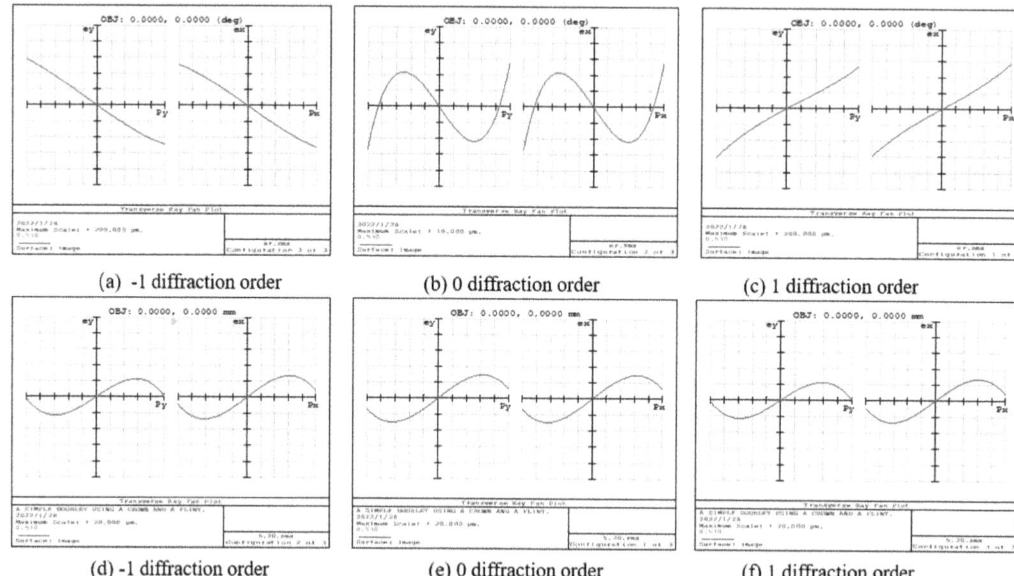

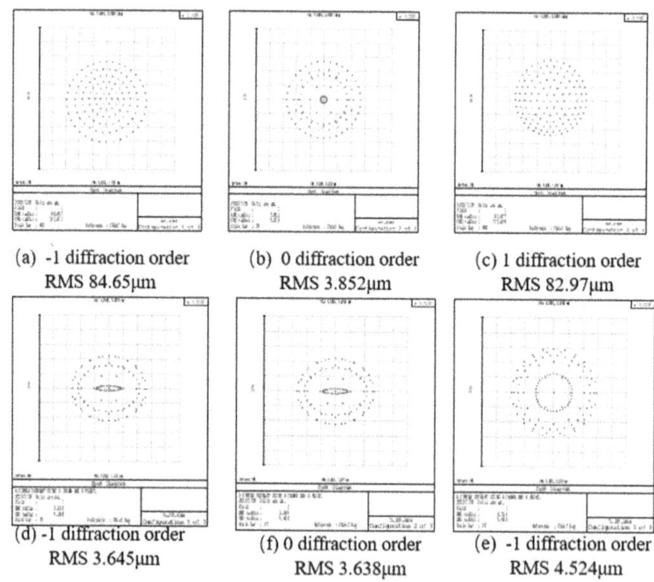

Figure 4. Light sectors calculated by ZEMAX in three orders for the 3D chromatography microscope system before and after aberration correction. (**a–c**) Light sectors corresponding to different diffraction levels of unmodified chromatography microscope system. (**d–f**) Light sectors corresponding to different diffraction levels of the aberrated chromatography microscope system.

Figure 5. Point column diagram in three orders for the 3D chromatography microscope system before and after aberration correction. (**a–c**) Point column diagram corresponding to different diffraction levels of unmodified chromatography microscopy system. (**d–f**) Point column diagram corresponding to different diffraction levels of aberrated chromatography microscopy system (compare RMS at the bottom of the diagram).

The system imaging quality is generally judged by comparing the radius of the diffuse spot in the point column diagram. The smaller the radius is, the better the imaging performance is. In Figure 5, the improved chromatography microscopy system shows a significant improvement in the diffusion range compared to the unimproved system. The diffuse spot radius of the improved laminar microscopy system is reduced by 80 μm for the ±1st diffraction order and by 0.2 μm for the 0th diffraction order. Thus, the improved off-axis Fresnel zone plate has the best performance of aberration correction.

The Seidel coefficients are an important tool for evaluating the imaging quality of a system. Each of the coefficients represents a different aberration in the imaging system. The smaller the absolute value of the coefficients is, the smaller the corresponding aberrations are. The Seidel coefficients and aberrations of the original 3D microscopy system based on the off-axis Fresnel zone plate and the improved system are shown in Figure 6. The results show that the aberrations in the chromatographic microscopy system are reduced after the optimization of the off-axis Fresnel zone plate according to the design proposed in this paper. The spherical aberration W040, coma aberration W131, astigmatism W222, and field curvature W220 are reduced by 62–70%, 96–98%, 71–82%, and 96–96% for the 0th and the ±1st diffraction light levels. This proves that the grid line distribution satisfies the expected results.

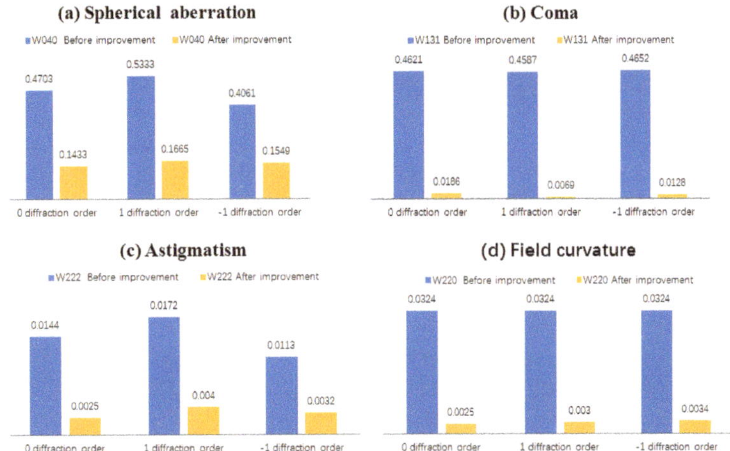

Figure 6. The comparison diagram of the Seidel coefficient of the tomographic microscopic system Blue corresponds to the Seidel coefficient for the unmodified chromatography system. The yellow color corresponds to the Seidel coefficient of the aberrated chromatography system. The adjacent blue and yellow rectangles correspond to one diffraction order of the chromatographic system.

2.2. The Groove Structure Design of the Aberrated Off-Axis Fresnel Zone Plate

The etching depth of the off-axis Fresnel zone plate determines the energy of each diffraction order. A 2nd level off-axis Fresnel zone plate is used, and its actual duty cycle is 1:1. In order to make the improved off-axis Fresnel zone plate satisfy the same diffraction efficiency for each order, the diffraction efficiency of the Fresnel zone plate is analyzed as a function of the etching depth using the software PCGrate. The intersection of the efficiency etching depth curves of 0th and ±1st diffraction order is the required etching depth, as shown in Figure 7. The energy of the first three diffraction levels is all 27.67% of the total incident light energy when $n = 1.46$ and the etching depth is 374 nm.

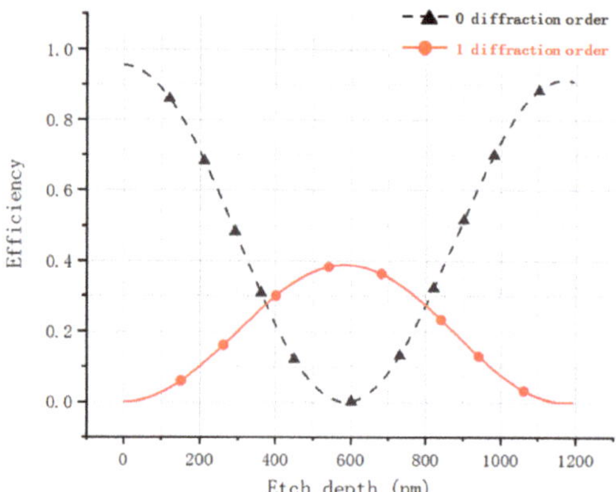

Figure 7. Diffraction efficiency variation curve with etching depth for 0 and ±1 diffraction levels of aberrated Fresnel zone plate (red represents level 0, black represents level ±1).

In this paper, the slot type of aberrated off-axis Fresnel zone plate used is a rectangular slot with a variable period; therefore, the effect of etching depth on its diffraction efficiency at different periods is discussed in this paper. It shows that the difference between the diffraction efficiency of 0th order and ±1st order is 0, and the diffraction efficiency does not vary with the period for a certain duty cycle with an etching depth of 374 ± 3 nm in Figure 8. The rectangular slot may not be achieved due to errors in the fabrication process. The difference between the diffraction efficiency of 0th order and ±1st order is 0 and the diffraction efficiency does not vary with the bottom angle of the slot type when the etching depth is 374 ± 3 nm at a certain period, as shown in Figure 8.

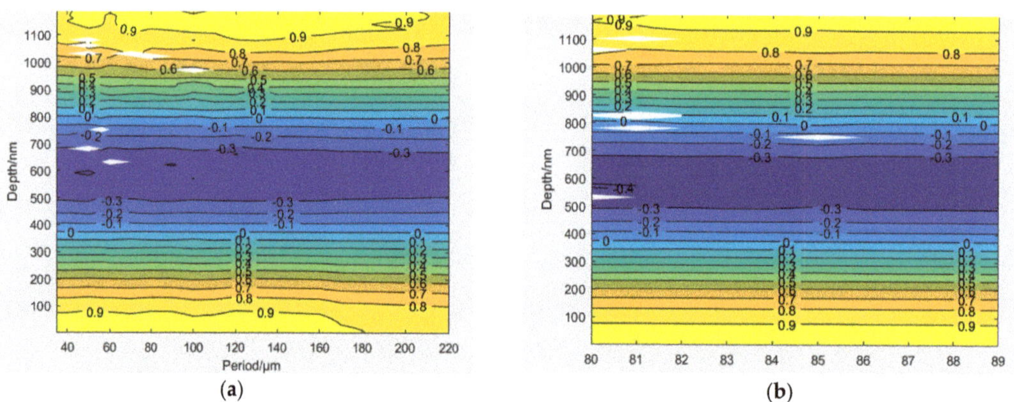

Figure 8. (**a**) Diffraction efficiency difference between level 0 and +1 diffraction versus period (μm) and etching depth (μm) contours (**b**) Diffraction efficiency difference between level 0 and +1 with respect to the contour of the bottom angle (°) and etching depth (μm).

3. Fabrication, Performance Test, and Simulation of Aberrated Off-Axis Fresnel Zone Plate

In this research, the wet etching process of amorphous material was used for the fabrication of an aberrated off-axis Fresnel zone plate. The main fabrication process includes substrate cleaning, homogenization, exposition, development, and wet etching [16]. First, the quartz substrate was cleaned. The quartz substrate is composed of quartz material (model 7980-0AA made by Corning, Corning, NY, USA), with a diameter of 76.2 mm and a thickness of 1.5 mm. The organic impurities and metal ions on the surface of the quartz substrate were cleaned using SC-1 liquid (H_2O: H_2O_2: NH_4OH = 5:1:1) and SC-2 liquid (H_2O:H_2O_2: HCl = 6:1:1), respectively. Then, the quartz substrate was cleaned in the order of toluene→acetone→alcohol→water. After cleaning, the substrate was pre-baked at 120 °C for 30 min, and then the quartz substrate was coated with a photoresist using the spin coating method. The photoresist model is BP212-7P, the spin coating speed was 3000 rpm, the coating time was 30 s, and the thickness of the photoresist was 500 nm. After the coating of the photoresist, the quartz substrate was placed in an oven for film hardening at 90 °C for 20 min. Next, the coated photoresist was exposed using UV mask contact exposure with an illumination level of 45 lux and an exposure time of 25 s. Immediately after the exposure, the photoresist was developed with a 3‰ NaOH solution and post-dried. Finally, the quartz was etched with a BHF solution (40% HF:49% NH3F = 1:8). To achieve 374 $\pm$ 3 nm grating slot fabrication, the etching time was 374 s because the BHF solution used in the experiment etches the quartz substrate at a rate of 1 nm/s.

After the fabrication of the aberrated off-axis Fresnel zone plate, the system was built after the performance test. The performance parameters of the aberrated off-axis Fresnel zone plate include the center period, slot shape, and diffraction efficiency. The central period and the slot pattern were measured using a Dimension Icon atomic force microscope with a probe placed at the center of the aberrated off-axis Fresnel zone plate to obtain the microstructure and parameters. The diffraction efficiency test of the aberrated off-axis Fresnel zone plate was carried out in two ways. First, the diffraction efficiency was calculated in the software program PCGrate based on the structural parameters acquired with the Dimension Icon microscope. Next, a single-wavelength off-axis Fresnel zone plate diffraction efficiency test optical system was built using a 516.5 nm laser source. The optical system mainly consists of a laser, an aberrated off-axis Fresnel zone plate, and a detector. Nine points (3 × 3) are measured uniformly on the Fresnel zone plate [17]. The ratio of the energy of each diffraction order to the laser energy was calculated.

After passing the performance test of the aberrated off-axis Fresnel zone plate, the laminar microscopy system was built; however, since the 3D profile of the geometrical optics does not visually show the imaging performance of the system, the optical system described above was simulated based on the physical optics of the VirtualLab Fusion software [18]. The simulation based on physical optics was performed by solving Maxwell's equations during the imaging process of the whole optical system to achieve field tracing; therefore, the simulation can be used to provide a more intuitive imaging result.

Finally, the optical components are selected to construct an adjustable cage system according to the optical system simulation in the software program ZEMAX, as shown in Figure 4. The microscope light source was white light, and a bandpass filter of 20 nm was taken. At the same time, the aberrated off-axis Fresnel zone plate was placed after two achromatic lenses with a focal length of 100 mm. The off-axis Fresnel zone plate and aberrated off-axis Fresnel zone plate were used for image acquisition experiments of the same specimen without adjustment of the instrument operating parameters or the distances of each optical element.

4. Result and Discussion

4.1. Performance Parameter Tests of Aberrated Off-Axis Fresnel Zone Plate

After the fabrication, the microscopic slot measurement was carried out on the aberrated off-axis Fresnel zone plate using an atomic force microscope. The results of the measurement are shown in Figure 9. The center period was 58 μm, the left edge period was 55 μm, and the right edge period was 40 μm, with a duty cycle of 0.5. The test results were in accordance with the expectation.

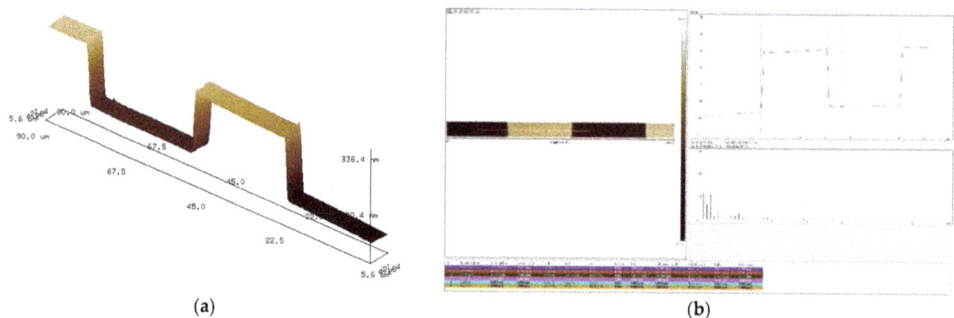

(a)

(b)

Figure 9. Test results of aberrated off-axis Fresnel zone plate. (**a**) the microstructure (**b**) the atomic force software tests.

In this paper, the diffraction efficiencies of 0th order and ±1st order in the central region of the aberrated off-axis Fresnel zone plate can be calculated using the software PCGrate with the parameters measured in the slot test. The diffraction efficiency result of the fabricated aberrated off-axis Fresnel zone plate is shown in Figure 10. The total diffraction efficiency of the three diffraction orders reaches 80%. It is 82.97% when the wavelength is 516.5 nm.

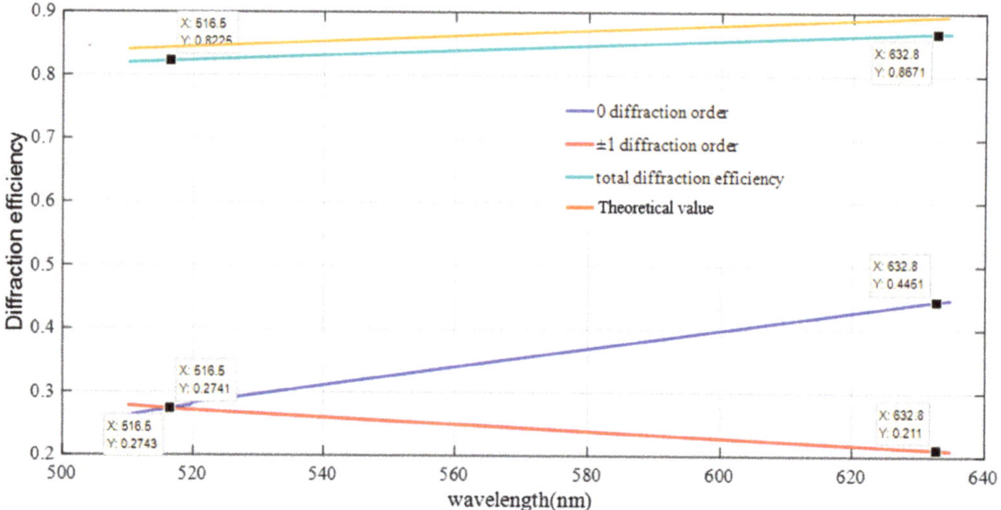

Figure 10. Inversion results of aberrated off-axis Fresnel wave with slice slot type test.

The diffraction efficiency test optical system of the single wavelength (516.5 nm) aberrated off-axis Fresnel zone plate is shown in Figure 11. The test was carried out by taking points on the aberrated off-axis Fresnel zone plate, as shown in Figure 11, and the test results are shown in Table 3. The average diffraction efficiency of the nine points on the aberrated off-axis Fresnel zone plate is 82.03%, which is in accordance with the expectation. A comparison between the measured and actual values of the diffraction efficiency for different diffraction orders is shown in Figure 12. The results from PCGrate are calculated with the ideal conditions of an aberrated off-axis Fresnel zone plate, which in practice causes errors with the theoretical values due to the influence of the surface roughness.

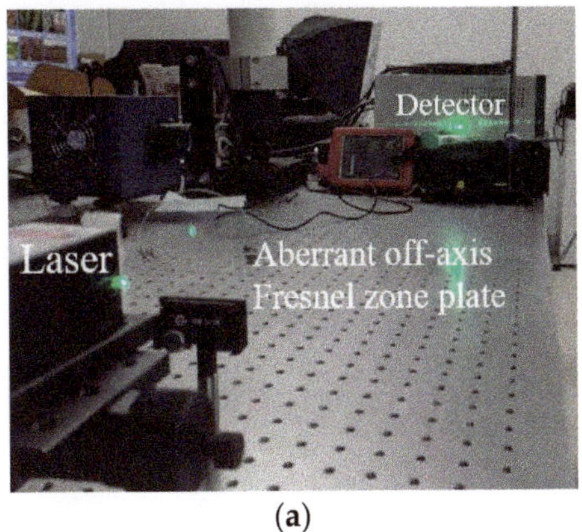
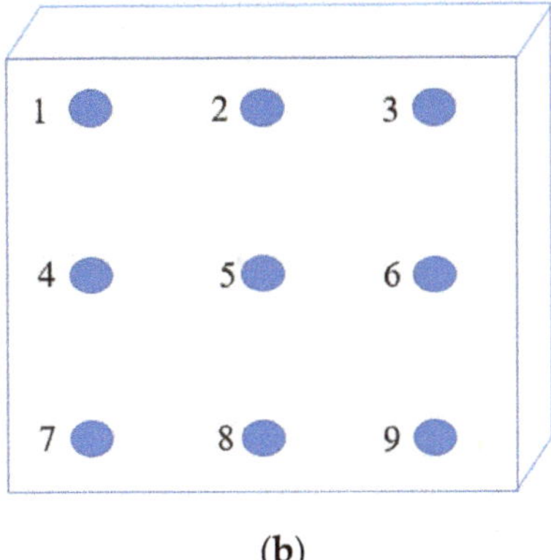

(**a**)　　　　　　　　　　　　(**b**)

Figure 11. (**a**) Single-wavelength diffraction efficiency test optical path. (**b**) Distribution of sampling points for aberrated off-axis Fresnel waveband diffraction efficiency test.

Table 3. Test results (diffraction efficiency).

No.	0 Level	+1 Level	−1 Level	Total
1	27.38%	27.30%	27.28%	81.96%
2	27.41%	27.35%	27.32%	82.08%
3	27.39%	27.33%	27.31%	82.03%
4	27.50%	27.25%	27.28%	82.03%
5	27.38%	27.31%	27.29%	81.98%
6	27.40%	27.38%	27.35%	82.13%
7	27.39%	27.34%	27.35%	82.08%
8	27.40%	27.33%	27.31%	82.04%
9	27.38%	27.31%	27.29%	81.98%

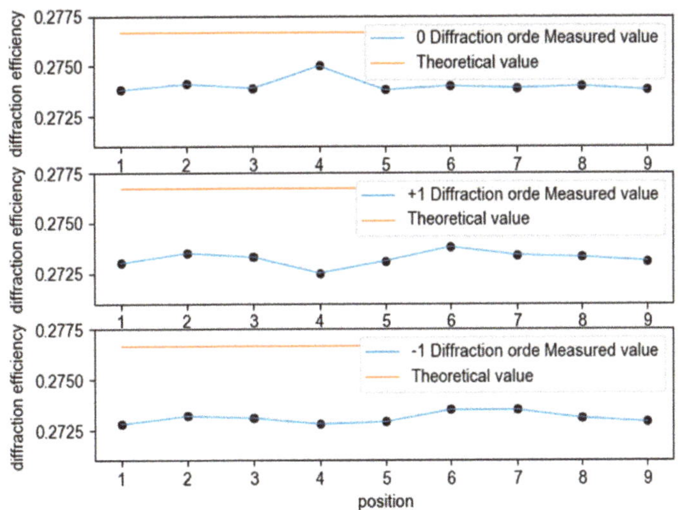

Figure 12. Comparison of theoretical and measured values of diffraction efficiency for different diffraction stages.

4.2. Imaging Experimental Analysis

The simulation results of the imaging in VirtualLab Fusion are shown in Figure 13. As shown in Figure 13, the images acquired after the improvement based on this method are significantly improved in terms of the accuracy of imaging and uniformity of intensity compared with the three objects.

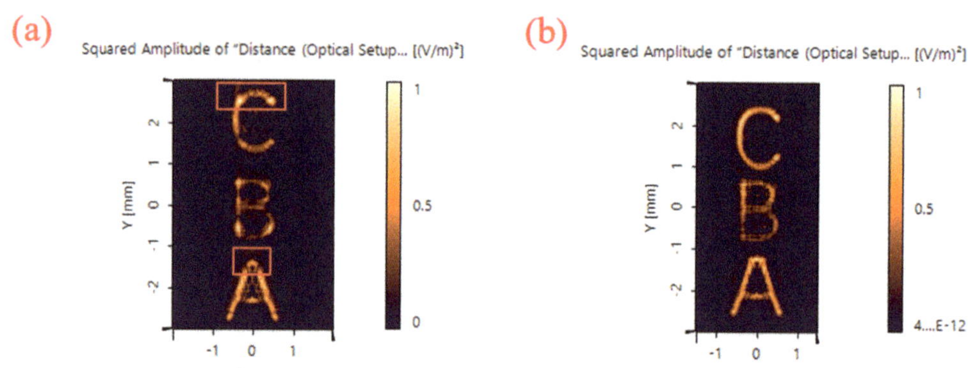

Figure 13. Imaging results: (**a**) Off-axis Fresnel zone plate; (**b**) aberrated off-axis Fresnel zone plate.

In terms of the image metrics, the signal-to-noise ratio and the average gradient were used to quantitatively evaluate the images, as shown in Table 4. The signal-to-noise ratio represents the ratio of the image signal to the noise. The average gradient represents the clarity of the image. The larger the average gradient is, the clearer the image is. Compared with the images with the unimproved chromatography microscope, the signal-to-noise ratio of the images acquired by the improved laminar microscope increases by 18%, and the average gradient increases by 3%.

Table 4. Laminar microscopy optical path image index.

Image Index	SNR [dB]	Average Gradient
Off-axis Fresnel zone plate	7.8143	6.0893
Aberrated off-axis Fresnel zone plate	9.3562	6.2719

The laminar microscopy optical system was built, as shown in Figure 14. With no adjustment of the instrument operating parameters and the distances of each optical element, the biological cell sections were subjected to image acquisition using the chromatographic microscopic light path and the achromatic chromatographic microscopic light path, respectively. When the magnification of the microscope objective was 5× and 10×, respectively, the acquired biological cell images are shown in Figures 15 and 16. Figures 15a,b and 16a,b show the images acquired by the unmodified chromatographic microscopy system. Figures 15c,d and 16c,d show the images acquired by the achromatic chromatography microscopy system. As can be seen from Figures 15 and 16, the images acquired by the unimproved chromatography microscopy system are blurred and have distortions at the edges. In contrast, the images obtained by the achromatic chromatography microscopy system were clearer and were able to obtain detailed information about the biological cells. Each image index is shown in Table 5. The quantitative analysis was performed using the same image metrics, such as signal-to-noise ratio and average gradient. From Table 5, it can be seen that the signal-to-noise ratio of the aberrated chromatography microscope system was improved by 18.28% on average compared with the off-axis Fresnel zone plate chromatography microscope system. The average gradient is improved by 2.85% on average; therefore, the results show that the imaging performance of the chromatography microscopy system has been improved by the design method of the off-axis Fresnel zone plate proposed in this paper. The feasibility of this method is fully verified by the experiments. Further, the deviation of the actual imaging results from the physical simulation results is mainly due to the fact that the three-layer object information is digital image information (400×400-pixel points) and single wavelength imaging during the simulation.

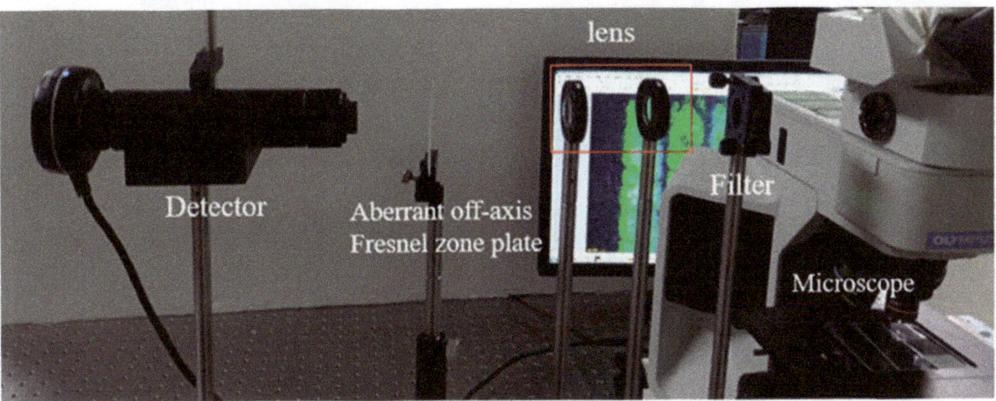

Figure 14. Experimental optical path diagram of chromatography microscope system.

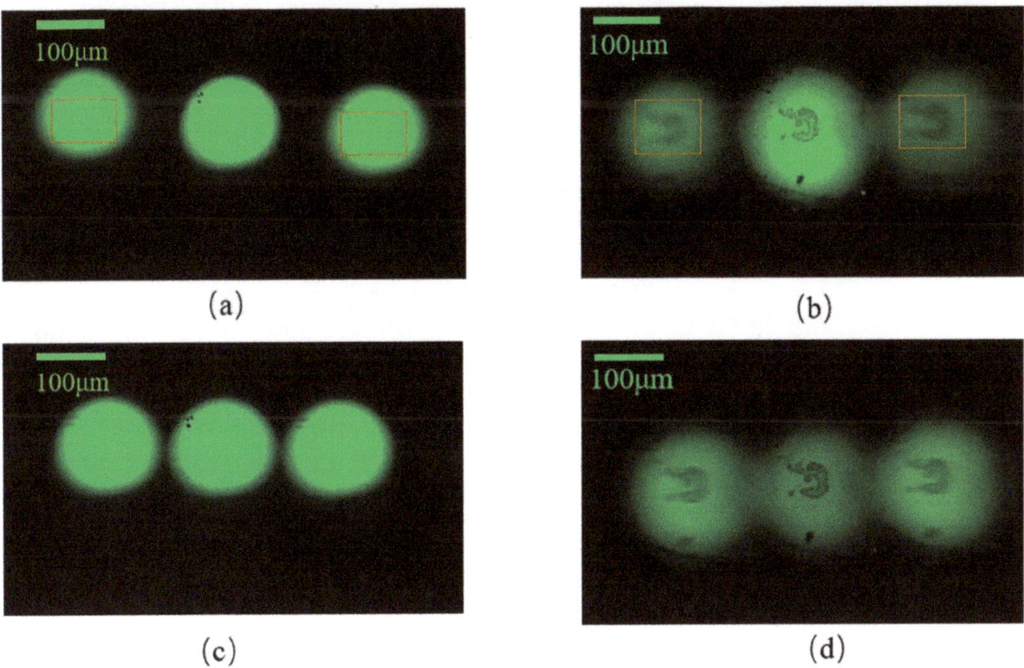

Figure 15. Results of experiment using chromatography microscopy with 5× objective lens with: (**a,b**) Off-axis Fresnel zone plate; (**c,d**) aberrated off-axis Fresnel zone plate.

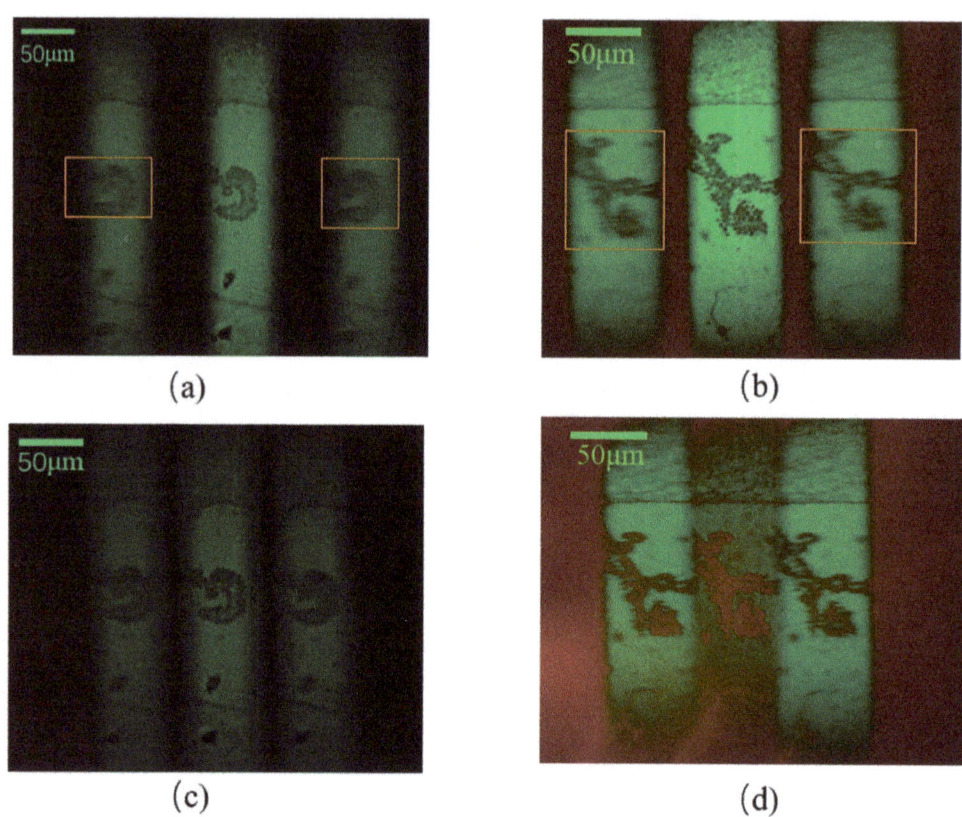

Figure 16. Results of experiment using chromatography microscopy with 10× objective lens with: (**a**,**b**) Off-axis Fresnel zone plate; (**c**,**d**) aberrated off-axis Fresnel zone plate.

Table 5. Laminar microscopy optical path image index.

Image Index	Off-Axis Fresnel Zone Plate	Aberrated Fresnel Zone Plate
SNR (5×) group 1	17.5556	20.8156
SNR (5×) group 2	21.5643	25.4458
SNR (10×) group 3	21.8550	25.8981
SNR (10×) group 4	21.7625	25.7106
Average gradient (5×) group 1	2.7128	2.7889
Average gradient (5×) group 2	2.7145	2.7905
Average gradient (10×) group 3	3.5359	3.6419
Average gradient (10×) group 4	2.7323	2.8096

5. Conclusions

This research is based on a holographic plane with interference fringes formed by two exposure points that can reduce the aberration of the system. The aberration coefficient calculation method based on the optical range function theory is used to design the aberration-eliminating off-axis Fresnel zone plate. The feasibility of the above method is verified by comparing a software simulation and an experiment. The simulated results show the reductions in the spherical aberration W040 by 62–70%, coma W131 by 96–98%, image dispersion W222 by 71–82%, and field curvature W220 by 96–96% after replacing the off-axis Fresnel zone plate with the aberration eliminating off-axis Fresnel zone plate. This is theoretically feasible. In the experimental aspect, the microscopic optical path was constructed, the cell images were acquired, and the image indexes were evaluated. The comparison of the image metrics shows that the average gradient increases by 2.8% and the signal-to-noise ratio increases by 18%.

In summary, the improved design of the off-axis Fresnel zone plate, which is the core element of the laminar microscope system, improves all types of geometric aberrations. Additionally, no post-processing of the images is required; therefore, the imaging quality of the chromatographic microscopy system is improved using this method.

Author Contributions: Conceptualization, L.Y., X.T. and Z.M.; Production of charts, L.Y., W.Z., J.P., J.Z., Y.Z. and Y.X.; Data collection:, L.Y., H.L., S.L. and Y.L.; Written, reviewed and edited, L.Y., X.T., Z.M. and Q.J. All authors have read and agreed to the published version of the manuscript.

Funding: Chinese Academy of Sciences Strategic Pioneering Science and Technology Special Project (XDA28050200).

Acknowledgments: Chinese Academy of Sciences Strategic Pioneering Science and Technology Special Project (XDA28050200); Jilin Province Science & Technology Development Program in China (20200403062SF) (20200401141GX) (20210201023GX) (20210201140GX) (20210203059SF); Chinese Academy of Sciences research instrumentation development project(YJKYYQ20200048); Science and Technology Innovation Platform of Jilin Province (20210502016ZP).

Conflicts of Interest: The authors declare no conflict of interest.

References

1. Wang, P.H.; Singh, V.R.; Wong, J.-M.; Sung, K.-B.; Luo, Y. Non-axial-scanning multifocal confocal microscopy with multiplexed volume holographic gratings. *Opt. Lett.* **2017**, *42*, 346–349. [CrossRef] [PubMed]
2. Blanchard, P.M.; Greenaway, A.H. Simultaneous Multiplane Imaging with a Distorted Diffraction Grating. *Appl. Opt.* **1999**, *38*, 6692–6699. [CrossRef] [PubMed]
3. Angarita-Jaimes, N.; McGhee, E.; Chennaoui, M.; Campbell, H.I.; Zhang, S.; Towers, C.E.; Greenaway, A.H.; Towers, D.P. Wavefront sensing for single view three-component three-dimensional flow velocimetry. *Exp. Fluids* **2006**, *41*, 881–891. [CrossRef]
4. Dalgarno, P.A.; Dalgarno, H.I.C.; Putoud, A.; Lambert, R.; Paterson, L.; Logan, D.C.; Towers, D.P.; Warburton, R.J.; Greenaway, A.H. Multiplane imaging and three dimensional nanoscale particle tracking in biological microscopy. *Opt. Express* **2010**, *18*, 877–884. [CrossRef]
5. Feng, Y.; Scholz, L.; Lee, D.; Dalgarno, H.; Foo, D.; Yang, L.; Lu, W.; Greenaway, A. Multi-mode microscopy using diffractive optical elements. *Eng. Rev.* **2011**, *31*, 133–139.
6. Feng, Y.; Dalgarno, P.A.; Lee, D.; Yang, Y.; Thomson, R.R.; Greenaway, A.H. Chromatically-corrected, high-efficiency, multi-colour, multi-plane 3D imaging. *Opt. Express* **2012**, *20*, 20705–20714. [CrossRef] [PubMed]
7. Abrahamsson, S.; Ilic, R.; Wisniewski, J.; Mehl, B.; Yu, L.; Chen, L.; Davanco, M.; Oudjedi, L.; Fiche, J.-B.; Hajj, B.; et al. Multifocus microscopy with precise color multi-phase diffractive optics applied in functional neuronal imaging. *Biomed. Opt. Express* **2016**, *7*, 855–869. [CrossRef] [PubMed]
8. He, K.; Wang, Z.; Huang, X.; Wang, X.; Yoo, S.; Ruiz, P.; Gdor, I.; Selewa, A.; Ferrier, N.J.; Schrerer, N.; et al. Computational multifocal microscopy. *Biomed. Opt. Express* **2018**, *9*, 6477–6496. [CrossRef] [PubMed]
9. Yuan, X.; Feng, S.; Nie, S.; Chang, C.; Ma, J.; Yuan, C. Multi-plane unequal interval imaging based on polarization multiplexing. *Opt. Commun.* **2019**, *431*, 126–130. [CrossRef]
10. Wolbromsky, L.; Dudaie, M.; Shinar, S.; Shaked, N.T. Multiplane imaging with extended field-of-view using a quadratically distorted grating. *Opt. Commun.* **2020**, *463*, 125399. [CrossRef]
11. Abrahamsson, S.; Chen, J.; Hajj, B.; Stallinga, S.; Katsov, A.Y.; Wisniewski, J.; Mizuguchi, G.; Soule, P.; Mueller, F.; Darzacq, C.D.; et al. Fast multicolor 3D imaging using aberration-corrected multifocus microscopy. *Nat. Methods* **2013**, *10*, 60–63. [CrossRef] [PubMed]
12. Liu, H.; Bailleul, J.; Simon, B.; Debailleul, M.; Colicchio, B.; Haeberlé, O. Tomographic diffractive microscopy and multiview profilometry with flexible aberration correction. *Appl. Opt.* **2014**, *53*, 748–755. [CrossRef] [PubMed]
13. Namioka, T.; Seya, M.; Noda, H. Design and Performance of Holographic Concave Gratings. *Jpn. J. Appl. Phys.* **1976**, *15*, 1181. [CrossRef]
14. Itou, M.; Harada, T.; Kita, T. Soft x-ray monochromator with a varied-space plane grating for synchrotron radiation: Design and evaluation. *Appl. Opt.* **1989**, *28*, 146–153. [CrossRef] [PubMed]
15. Harada, T.; Moriyama, S.; Kita, T. Mechanically ruled stigmatic concave gratings. *Jpn. J. Appl. Phys.* **1975**, *14*, 175. [CrossRef]

16. Jiao, Q.; Zhu, C.; Tan, X.; Qi, X.; Bayanheshig. The effect of ultrasonic vibration and surfactant additive on fabrication of 53.5 gr/mm silicon echelle grating with low surface roughness in alkaline KOH solution. *Ultrason. Sonochemistry* **2018**, *40*, 937–943. [CrossRef] [PubMed]
17. Palmer, E.W.; Hutley, M.C.; Franks, A.; Verrill, J.F.; Gale, B. Diffraction gratings (manufacture). *Rep. Prog. Phys.* **2001**, *38*, 975. [CrossRef]
18. Zhang, B.; Rao, P.; Fuso, F.; Guiying, H.; File, W.D. Simulation of optical distribution inside of SNOM optical probes with VirtualLab Fusion. In Proceedings of the Laser Ignition Conference, Bucharest Romania, 20–23 June 2017; pp. 17–22.

Communication

A Fiber-Integrated CRDS Sensor for In-Situ Measurement of Dissolved Carbon Dioxide in Seawater

Mai Hu [1,2], Bing Chen [1], Lu Yao [1], Chenguang Yang [3], Xiang Chen [1] and Ruifeng Kan [1,*

[1] Anhui Institute of Optics and Fine Mechanics, Hefei Institute of Physical Science, Chinese Academy of Sciences, Hefei 230031, China; humai@aiofm.ac.cn (M.H.); bchen@aiofm.ac.cn (B.C.); lyao@aiofm.ac.cn (L.Y.); xchen@aiofm.ac.cn (X.C.)
[2] University of Science and Technology of China, Hefei 230026, China
[3] Institute of Deep-Sea Science and Engineering, Chinese Academy of Sciences, Sanya 572022, China; yangcg@idsse.ac.cn
* Correspondence: kanruifeng@aiofm.ac.cn; Tel.: +86-0551-65593695

Abstract: Research on carbon dioxide (CO_2) geological and biogeochemical cycles in the ocean is important to support the geoscience study. Continuous in-situ measurement of dissolved CO_2 is critically needed. However, the time and spatial resolution are being restricted due to the challenges of very high submarine pressure and quite low efficiency in water-gas separation, which, therefore, are emerging the main barriers to deep sea investigation. We develop a fiber-integrated sensor based on cavity ring-down spectroscopy for in-situ CO_2 measurement. Furthermore, a fast concentration retrieval model using exponential fit is proposed at non-equilibrium condition. The in-situ dissolved CO_2 measurement achieves 10 times faster than conventional methods, where an equilibrium condition is needed. As a proof of principle, near-coast in-situ CO_2 measurement was implemented in Sanya City, Haina, China, obtaining an effective dissolved CO_2 concentration of ~950 ppm. The experimental results prove the feasibly for fast dissolved gas measurement, which would benefit the ocean investigation with more detailed scientific data.

Keywords: seawater dissolved gas; carbon dioxide; optical cavity ring-down spectroscopy; in-situ measurement

Citation: Hu, M.; Chen, B.; Yao, L.; Yang, C.; Chen, X.; Kan, R. A Fiber-Integrated CRDS Sensor for In-Situ Measurement of Dissolved Carbon Dioxide in Seawater. *Sensors* **2021**, *21*, 6436. https://doi.org/10.3390/s21196436

Academic Editor: Jin Li

Received: 21 August 2021
Accepted: 24 September 2021
Published: 27 September 2021

1. Introduction

Marine carbon cycling is the result of a series of physical, geological and biological processes on a spatiotemporal scale [1,2]. The ocean absorbs one-third of the anthropogenic carbon emission, about 2 billion tons per year. As such, the ocean becomes one important place for carbon sequestration [3]. Furthermore, CO_2 is the main greenhouse gas, the main dissolved gas of seawater and the main fluid component of the deep-sea extreme window of cold spring and hydrothermal solution. Precise measurement on its spatiotemporal distribution is significant to investigate the biogeochemical material cycle and global climate change [4,5]. However, common methods based on sampling-laboratory analysis are not enough to support modern marine science. In-situ CO_2 sensors with high sensitivity, high fidelity, large dynamic range and fast response are highly needed in many cutting-edge research topics, such as the sea-air exchange flux of CO_2 [6], CO_2 concentration of deep-sea cold spring and hydrothermal fluid components [7–9], and isotope measurement [7,8,10,11].

Currently, optical technology, semiconductor gas sensing and mass spectrometry [8,9,11–13] are common methods in in-situ measurement of dissolved gas in seawater. Among them, laser-based in-situ optical spectrometer is suitable for greenhouse gas sensing in seawater due to its unique selectivity and sensitivity [12,14,15]. Using infrared spectroscopy technology, the German Hydro C company demonstrated dissolved CH_4 measurement in seawater [16]. Using the off-axis integrated cavity output spectroscopy (OA-ICOS), the LGR Company in the United States measured the dissolved CH_4/CO_2 and its isotope $\delta^{13}CH_4$ in seawater [11]. Using mid-infrared absorption spectroscopy technology, Zheng Chuantao et al. achieved the measurement of dissolved CO_2 and its isotope $\delta^{13}CO_2$ [12]. In addition, cavity ring-down spectroscopy (CRDS), proposed by O, Keefe and Deacon in 1988 [17], has ultra-high detection sensitivity, light intensity jitter immunity and free instrument calibration, making CRDS one of the best candidates for gas detection. The past years have witnessed its remarkable progresses in precision spectroscopy [18–20] and important applications in atmospheric trace gas measurement [21–23]. However, the application of this technology in the marine field for in-situ measurement remains unresolved due to the challenges of high stability resonant cavity, high precision/low power circuit and time-consuming dissolved gas concentration retrieval.

In this paper, we report the development of an in-situ CRDS based dissolved CO_2 sensor. A fast exponential regression model is proposed to retrieve the concentration of dissolved gas in seawater. In the implementation, we design high-pressure assembling and use polydimethylsiloxane (PDMS) membrane for water/gas separation and enrichment [24,25]. A long-time in-situ observation near the coast is carried out to prove the feasibility of in-situ separation, enrichment and measurement of dissolved CO_2 in seawater.

2. Principle of CRDS-Based Seawater Dissolved Gas Measurement

2.1. Water/Gas Separation and Enrichment, and Dissolved Gas Retrieval

A PDMS membrane is one common tool, as the gas-liquid interface with a thickness of l, to separate and enrich seawater dissolved gases [26,27]. The concentration difference between both sides enables the dissolved gas pass through the membrane and blocks liquid water molecule $(H_2O)_n$. Thus, small gas molecules, such as CH_4, CO_2 and O_2 can be separated from seawater. The water/gas separation of PDMS membrane is typically described by the "dissolution-diffusion" model [24,26]. When concentration difference between both sides exists, gas molecules diffuse into the membrane and realize gas exchange.

In the case of a stable situation, the dissolved gas concentration remains stable along the direction of film thickness. The diffusion flux on the side of the gas chamber can be expressed by Fick's first law [24,26,27]:

$$F_G = \frac{D_G S_G A (P_{G1} - P_{G2})}{l}, \tag{1}$$

where F_G ($cm^3 \cdot cm^2 (cm^2 polymer)^{-1} \cdot s^{-1}$) is the diffusion flux of gas component G per unit time, D_G ($cm^2 \cdot s^{-1}$) is the diffusion coefficient of gas component G in the membrane, S_G ($cm^3 \cdot (cm^2 polymer)^{-1} \cdot Pa^{-1}$) is the solubility coefficient of gas component G in the membrane, P_{G1} (Pa) is the partial pressure of seawater dissolved gas G, P_{G2} (Pa) is the partial pressure of gas component G in the gas chamber, A (cm^2) is the film area, l (cm) is the film thickness. While gas diffusion flux can be expressed by Fick's second law in the case of unstable situation [25,26]:

$$F_{G,t} = F_{G,ss} \left(1 + 2 \sum_{n=1}^{\infty} (-1)^n \exp \left\{ \frac{-n^2 \pi^2 D_G t}{l^2} \right\} \right), \tag{2}$$

where, $F_{G,t}$ is the gas flux at time t, $F_{G,ss}$ is the gas flux in the stable situation.

The diffusion coefficient of CO_2 in PDMS membrane is about 1.5×10^{-5} cm^2/s, $l \ll 1$ cm, $t = 1$ s, then $\exp \left\{ \frac{-n^2 \pi^2 D_G t}{l^2} \right\}$ is almost zero and $F_{G,t} \approx F_{G,ss}$. Thus, the diffusion flux in this case can also be described by Fick's first law. Therefore, the concentration change of the gas component G is as follows:

$$P_{G2} * V = \int \frac{D_G S_G A (P_{G1} - P_{G2})}{l} d_t, \tag{3}$$

where V is the volume of the gas chamber. From Equation (3), we can obtain:

$$\frac{dP_{G2}}{dt} * V = \frac{D_G S_G A (P_{G1} - P_{G2})}{l},\tag{4}$$

With the boundary condition, $t \rightarrow \infty$, $P_{G2} = P_{G1}$, it will be extrapolated that:

$$P_{G2} = K * \exp\left(-\frac{D_G S_G A}{lV}t\right) + P_{G1},\tag{5}$$

where K is related to the initial pressure. For non-condensable gases, such as CO_2 and CH_4, the value of $-\frac{D_G S_G A}{lV}$ is independent from partial pressure.

After measuring the value of P_{G2} over time in an unbalanced situation, the exponential regression is used to retrieve P_{G1}. P_{G1} equals to P_{G2} in balanced situation.

2.2. Optical Measurement Using CRDS

When laser with an intensity of I_{in} passes through uniform gas substance, the laser decays to I_{out} due to gas absorption, which can be described by the Beer-Lamber law [10,17]:

$$I_{out} = I_{in} * \exp(-\alpha * L),\tag{6}$$

where, α (cm^{-1}) is the spectral absorption coefficient, L (cm) is the interaction distance.

In the configuration of CRDS, a couple of high reflectivity mirrors, usually higher than 99.99%, enable a significant interaction length extension inside a limited physical space [28], and become capable of detecting minor absorption of trace gas concentration. Its working principle is shown in Figure 1. When laser beam resonates with one resonant cavity mode, the laser power inside the cavity will be rapidly built up. With a trigger that the inside laser power reaches a certain threshold, the incident laser is quickly cut off to generate a free intracavity ring down. The leaked laser intensity $I_v(t)$ after each ring down is successively recorded at the exit to obtain optical intensity that decays with time as follows,

$$I_v(t) = I_{0v} * \exp\left(-\frac{tc}{L_{mirrors}}(1 - R + \alpha_v L_{mirrors})\right),\tag{7}$$

where, c is the speed of light, t (µs) is the time, $L_{mirrors}$ (cm) is the physical distance between the two mirrors, R is the reflectivity of the optical cavity mirror, α_v (cm^{-1}) is the spectral absorption coefficient of the specific wavelength; I_{0v} is the initial laser intensity when the laser is cut off.

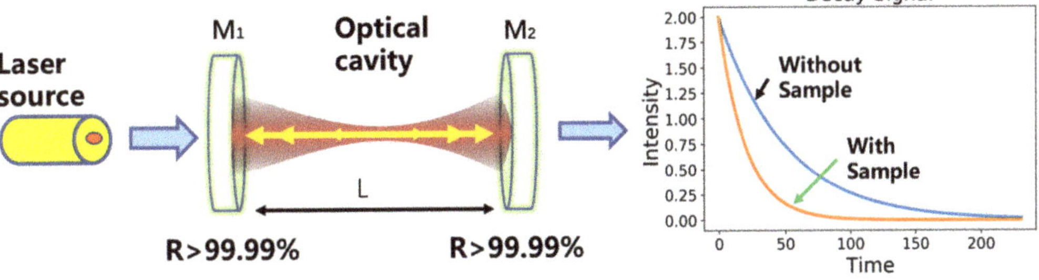

Figure 1. Schematic of the cavity ringdown technique. M_1 and M_2: cavity mirrors with high reflectivity (>99.99%).

The time for the initial laser intensity reduces to $1/e$ in the measurement is defined as the ring-down time [17]. According to Equation (7), in the case of no gas absorption, the empty cavity ring-down time τ_0 is:

$$\tau_0 = \frac{L_{mirrors}}{C(1-R)},\tag{8}$$

In the presence of gas absorption, the ring-down time τ_ν is:

$$\tau_\nu = \frac{L_{mirrors}}{C(1-R+\alpha_\nu L_{mirrors})},\tag{9}$$

Combining (8) and (9), we obtain the intracavity spectral absorption coefficient as:

$$\alpha_\nu = \frac{1}{C\tau_\nu} - \frac{1}{C\tau_0},\tag{10}$$

In the absorption spectrum [29], the absorption coefficient can be expressed as:

$$\alpha_\nu = S(T) * P_{total} * X * \psi(\nu),\tag{11}$$

where $S(T)$ ($cm^{-2}\cdot atm^{-1}$) is the intensity of the absorption line, P_{total} (atm) is the total pressure of the gas, X is the molecular concentration, $\psi(\nu)$ (cm) is the absorption line shape function.

Since integral of $\psi(\nu)$ over ν is equals 1, i.e., $\int_{-\infty}^{+\infty} \psi(\nu)d_\nu = 1$, $S(T)$ relates to the temperature for a specific absorption line, and P_{total} can be measured using a commercial pressure sensor. Therefore, after fitting the absorption coefficient curve to obtain the integrated area A, the partial pressure of the gas can be calculated by:

$$P = XP_{total} = \frac{A}{S(T)},\tag{12}$$

Thus, the partial pressure of dissolved gas (P_{G1}) can be retrieved by measuring the partial pressure of gas (P_{G2}) in the cavity and fitting formula (5).

3. In-Situ Dissolved CO_2 Sensor Configuration

Figure 2 depicts the schematic diagram of the optical dissolved-CO_2 sensor, which comprises three parts, one pressured chamber, one water/gas separation and enrichment unit and one gas measurement unit. The first part, i.e., the pressure chamber, is a dry titanium alloy chamber with an inner diameter of 128 mm, a length of 750 mm and a wall thickness of 10 mm. It can withstand a pressure as high as 57 MPa, which is suitable for experiments at about 4500 m under water. The second part, i.e., the water gas separation and enrichment unit consists of a water pump (SEA-BIRD SBE-5T), a PDMS membrane module (membrane thickness 50 μm, diameter 5 cm, gas separation efficiency 0.034 mL/min at 296 K and 1 atm pressure difference), a drying box, an air pump (KNF NMP05), a filter (Swagelok, SS-2F-05), two needle valves (Swagelok, SS-ORS2) and the gas chamber (optical resonant cavity). With the water pump, the seawater continuously flows through the surface of PDMS membrane at a flow rate of 0.8 L/min, forming a stable dissolved gas concentration field on the surface of the membrane. The dissolved gas permeates into the PDMS membrane. On the other side of the membrane, the permeated gas is desiccated by the drying box, then enters the gas chamber through the needle valve 1 and the filter, and finally returns to the PDMS membrane module through the filter, the needle valve 2 and the gas pump. Thus, water/gas separation and enrichment can be completed. The needle valve permits a flow rate of approximately 50 mL/min.

The third part, i.e., the gas detection unit, includes a control circuit (DSP, TMS320C6748, Texas Instruments, Dallas, TX, USA), a DFB laser (NLK1L5GAAA, NEL, Yokohama, Japan), a semiconductor optical amplifier (SOA, BOA1080P, Throlabs, USA), an isolator (PIISO-

1600-D-L-05-FA, Qinghe Photonics, Shenzhen, China), an optical resonator and a photo-electric detector (GPD, GAP1000FC, GPD Optoelectronics, Salem, MA, USA). Sawtooth signal and step signal generated by the control circuit are combined to drive the laser to achieve laser beam resonance enhancement inside the optical resonator. When the enhanced intracavity laser power, monitored by the detector, exceeds a certain threshold, the incident laser is rapidly cut off using the TTL driven SOA to generate ring-down signal. With the ring down signal recorded, a fast single exponential fitting is performed to calculate the ring down time. After 20 recordings of ring-down time for a single longitudinal mode, the step voltage is reset to match the laser to next longitudinal mode until the whole absorption line of CO_2 is covered. With the averaged absorption spectra, the CO_2 absorbance and concentration are calculated by spectral fitting algorithm. Finally, the concentration of the seawater side gas is retrieved according to the exponential fitting in the unbalanced situation.

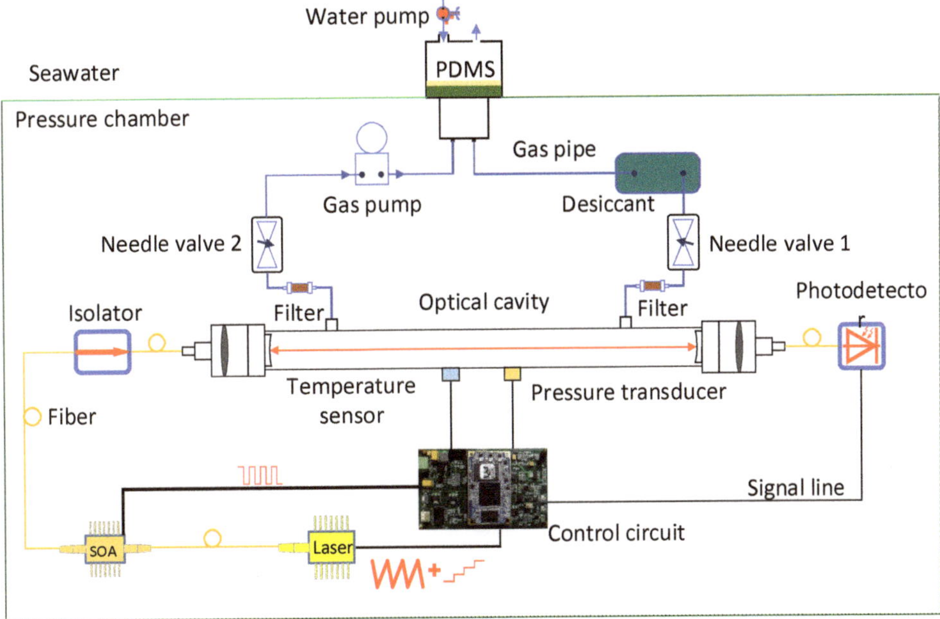

Figure 2. Framework of in-situ underwater instrument for dissolved gas measurement.

3.1. Absorption Line Selection

Appropriate absorption line with high absorption intensity and less background interference can benefit the dissolved CO_2 measurement with high signal-to-noise ratio. Considering the application condition in seawater dissolved gas measurement, absorption spectra of CO_2, $^{13}CO_2$, H_2O and CH_4 within 1599.4–1599.7 nm is simulated based on the HITRAN database [30] (temperature: 296 K, gas pressure: 1.01×10^5 Pa, CO_2 = 400 ppm, $^{13}CO_2$ =, H_2O = 2% and CH_4 = 2 ppm). The simulation results, shown in Figure 3, illustrate that CO_2 absorption coefficient is 2.5×10^{-7} cm^{-1} with negligible interference. Thus absorption transition R(36) at 6251.761 cm^{-1} is selected for the following CO_2 measurement.

3.2. Optical Resonator Design

The optical resonator is shown in Figure 4. Two identical cavity mirrors M1 and M2 (Layertec) have a diameter of 12.7 mm, a radius of curvature of 1000 mm and a reflectivity of higher than 99.99% (@1500~1700 nm). Lens1 and Lens2 are adjustable focus aspherical collimators (CFC-8X-C, Throlabs, Newton, NJ, USA). To suppress the high-order modes, the laser beam is spatially filtered using a single-mode fiber before illuminating on the photodetector, achieving a high-order modes suppression ratio of better than 100:1. It should be noted that, most of the optical and mechanical components of the optical cavity are fixed with structural adhesive to adapt to the extreme marine environment and improve the system stability.

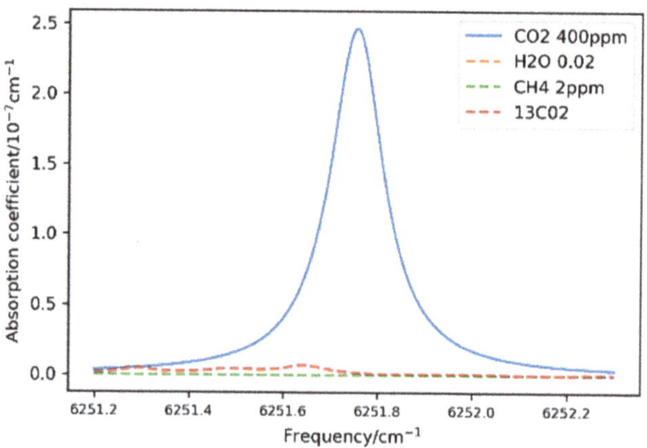

Figure 3. Simulation of CO_2 absorption feature and background interference from of $^{13}CO_2$, H_2O vapor and CH_4.

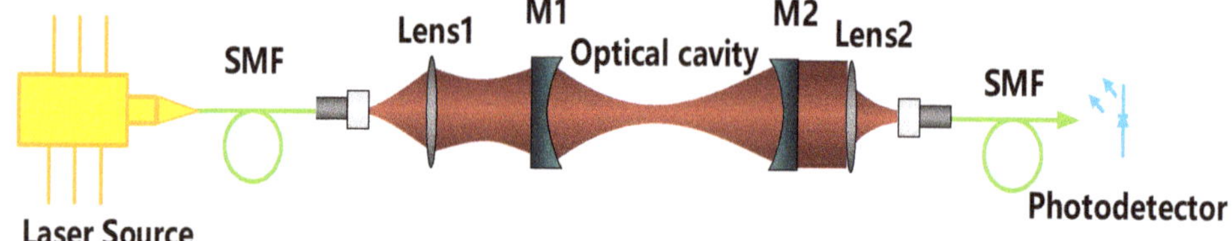

Figure 4. Framework of the optical resonant cavity, SMF: Single mode fiber; M1 and M2: cavity mirrors. Lens1 is a resonant cavity mode matching lens for mode matching of a laser-beam to a high finesse optical cavity. Lens2 is the light converging lens for coupling light into fiber.

3.3. Longitudinal Mode Matching and Spectral Scanning

The physical distance between the two cavity mirrors is 420 mm, corresponding to a free spectral range (FSR) of 0.0119 cm^{-1}. For each measurement using the continuous-wave CRDS system, the laser frequency is adjusted to resonate with one cavity mode. Since the cavity length remains stable during the measurement, the longitudinal cavity modes are used as the frequency reference to depict spectral absorption curve. To avoid potential spectral distortion from the longitudinal mode leakage during the spectral scanning process, we propose a strategy of longitudinal mode matching in a stepwise manner shown in Figure 5, and the step size is set as 1/5 FSR. Ideally, when the laser resonates with the longitudinal cavity mode q, 5 steps can rematch laser and the resonant cavity. However, it can be hardly realized due to the laser wavelength drift and cavity vibration. Differently, a high frequency sawtooth modulation (amplitude, $1/4FSR \leq M \leq 2/5FSR$) is simultaneously superimposed on the step signal to ensure the resonance of each cavity mode with the laser. In the implementation of this strategy, each longitudinal mode resonates at least once in 5-step scanning, and at least one step does not resonate. The discrete spectrum can be obtained by taking the non-resonant step as a marker.

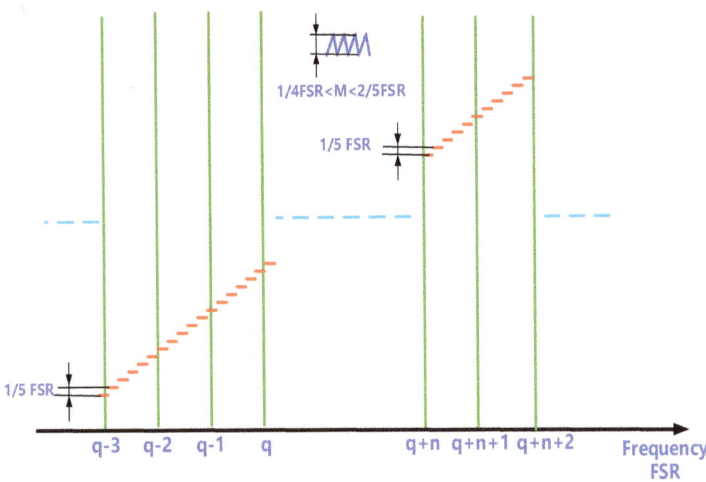

Figure 5. Schematic of the stepwise spectral scanning.

4. Experimental Results

4.1. Sensor Performance

After assembling the sensor, calibrated CO_2 sample with a certain concentration of 885 ppm (uncertainty, 1%) was sealed inside the gas chamber. The absorption spectrum of CO_2 is obtained using the method described in Section 3.3. Voigt spectrum fitting was performed using a Python LMFIT-based program. Figure 6 presents a direct comparison between raw spectrum data of and the fitting curve, and a residual error of 1.5×10^{-9} cm^{-1}. The SNR is calculated to be 500, corresponding to a detection limit of 1.8 ppm.

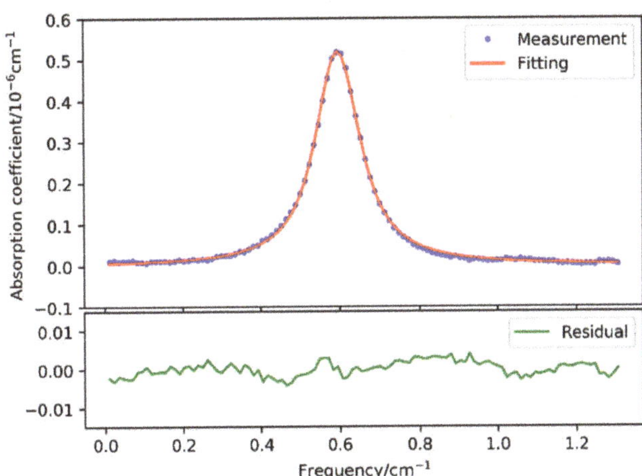

Figure 6. Comparison of the measured CO_2 spectral data and Voigt fitting curve.

CO_2 samples with different concentration of 2000 ppm, 1600 ppm, 1000 ppm, 500 ppm and 250 ppm were generated by diluting the calibrated 10000 ppm CO_2 with pure N_2. The uncertainty for above samples is 1%, mainly introduced by the dilution system. The mass flow meter was used to control the inlet flow rate at 50 mL/min. A continuous measurement of each CO_2 sample was performed over 45 min and the experimental results are shown in Figure 7a.

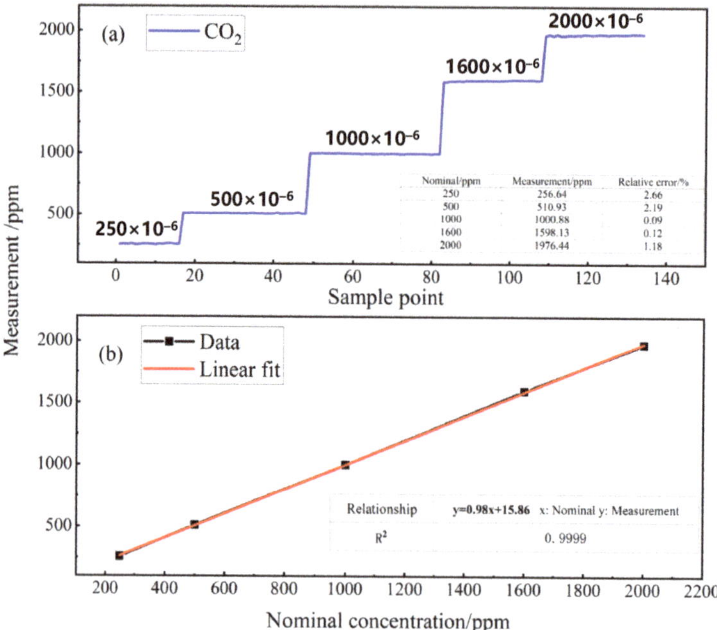

Nominal/ppm	Measurement/ppm	Relative error/%
250	256.64	2.66
500	510.93	2.19
1000	1000.88	0.09
1600	1598.13	0.12
2000	1976.44	1.18

Relationship	y=0.98x+15.86 x: Nominal y: Measurement
R^2	0.9999

Figure 7. (**a**) Measured results of CO$_2$ samples ranging from 250 ppm to 2000 ppm and (**b**) a linear fit to the measured data.

A linear fit, shown in Figure 7b, was thus performed to the measured data and the R-square value was calculated to be 0.9999, illustrating a good linear response to the CO$_2$ concentration from 250 ppm to 2000 ppm. The relative error of the measured values of the five groups of standard gases is less than 2.66%, which is consistent with the uncertainty of the standard gases.

To prove the feasibility, a long-term (8 h) comparison experiment was carried out between this sensor and one commercial land-based instrument CRDS instrument (G2201-i, Picarro). The gas pipelines of G2201-i and this sensor were connected together to guarantee the same the analyte was simultaneously and separately measured. The comparison result, shown in Figure 8, demonstrates that the measured CO$_2$ concentration and its variation trend are consistent with each other and the maximum relative difference is within 1.3%.

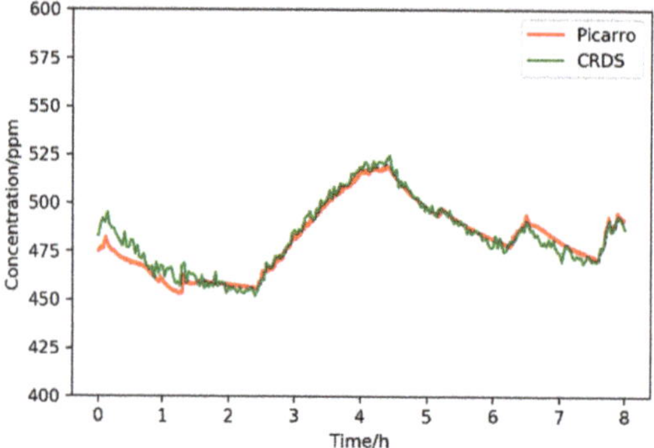

Figure 8. Comparison of this sensor to one commercial CRDS instrument (G2201-i, Picarro, Santa Clara, CA, USA).

4.2. Verification of Exponential Fit Retrieval Method for Dissolved CO_2 Concentration

Sample solutions are prepared by mixing 2000 ppm CO_2 and pure N_2 (purity: 99.99%) using two mass flow meters (AST10-DLCMX-100C-025-A2B2-4VE, Asert, Franklin, MA, USA), as shown in Figure 9a. A submersible pump (SBE-5T, SEA-BIRD, Bellevue, WA, USA) is used to continuously cycle the solution through the PDMS module to separate dissolved CO_2 to be measured. PDMS module is connected to the chamber of in-situ measurement system via a stainless tube. After 12-h measurement, the exponential model combined with Levenberg Marquardt (LM) algorithm is performed on the measured data. Figure 9b depicts the results with a R-square of 99.84%, when gas-phase CO_2 concentration in the chamber is lower than the dissolved CO_2. Figure 9c depicts the results with a R-square of 98.71% when gas-phase CO_2 concentration in the chamber is higher than the dissolved CO_2. The coefficient $\frac{D_G S_G A}{IV}$, independent from the gas concentration, inside and outside the membrane are 0.1389 and 0.1444, respectively. The coefficients of two different situations are consistent with each other with a small difference of 3.8%. Thus, the method of exponential fit for dissolved gas concentration proves to be validated.

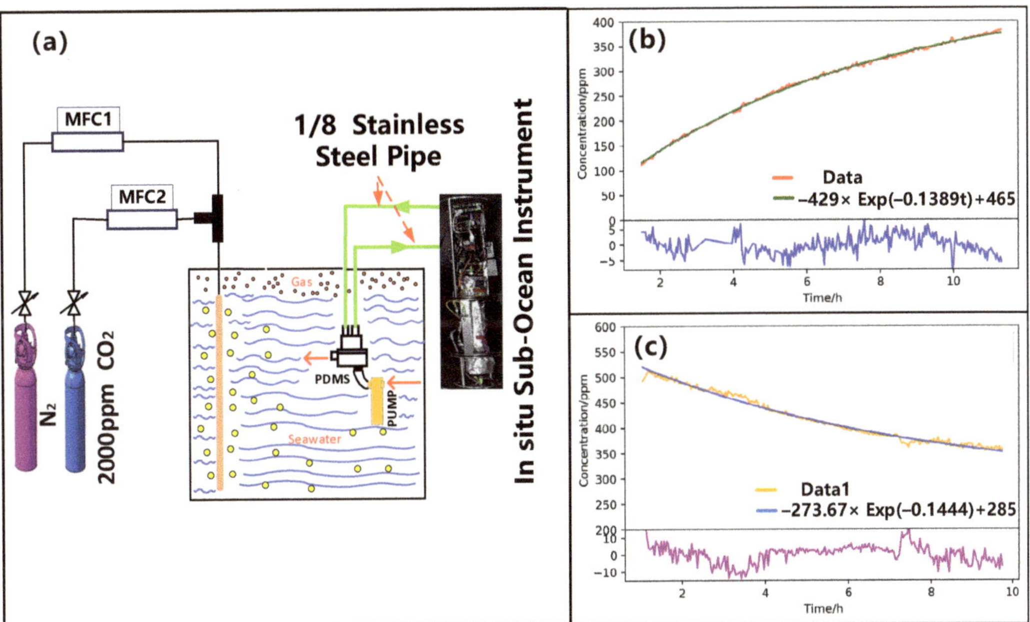

Figure 9. (**a**) Schematic of sample solution preparation and dissolved CO_2 measurement system. Real-time concentration measurements when CO_2 concentration in the cavity is (**b**) lower and (**c**) higher than the dissolved CO_2, respectively.

4.3. In-Situ Detection of Dissolved CO_2 in Seawater

From 6 to 7 March 2021, an in-situ observation was carried out near the coast of Sanya Institute of Deep Sea, Chinese Academy of Sciences, Sanya, Hainan Province. Figure 10 shows the observation location.

At the beginning of measurement, about 1100 ppm CO_2 was filled in the cavity to quickly balance the concentration inside and outside the membrane. Figure 11 shows the observation result, the whole measurement is divided into two unstable parts and one stable part. In the first unstable part (3 h), the measured concentration was higher than that of seawater dissolved CO_2 due to the presence of 1100 ppm CO_2 in the measurement chamber, and the CO_2 diffused from the measurement chamber into seawater. The concentration of seawater dissolved CO_2 is calculated to be 850 ppm. In the second stable part (9 h), the gas concentration inside and outside the membrane remained equal, and the dissolved gas concentration fluctuated between 950 ppm to 980 ppm. In the third unstable part (8 h), the concentration of dissolved CO_2 decreased and the CO_2 in the cavity continued to diffuse into seawater. The concentration of dissolved CO_2 in the water is calculated to be 808 ppm. Interestingly, these measured dissolved CO_2 concentrations are much higher than the atmospheric CO_2 concentration, about 400 ppm, mainly because of a number of yachts cruising near the coast and the tide phenomenon.

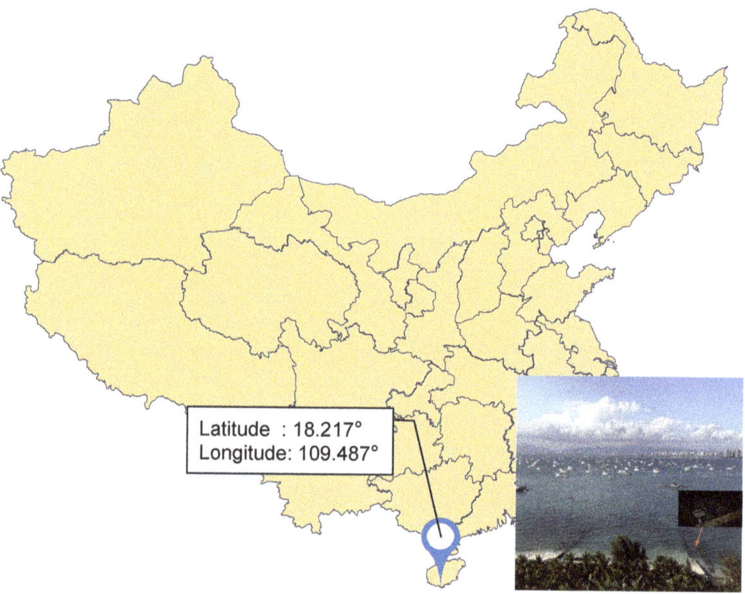

Figure 10. The in-situ test site is located at 18.217° north latitude and 109.487° east longitude.

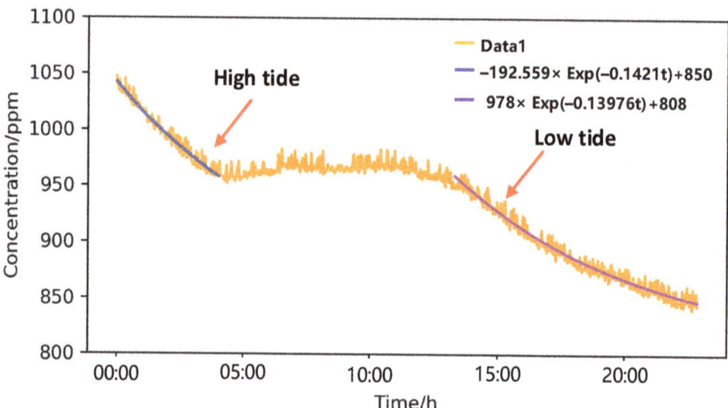

Figure 11. The result of In-situ measurements near coast.

5. Conclusions

We develop a fiber-integrated in-situ dissolved CO_2 sensor using CRDS, in which a PDMS membrane is employed for water/gas separation and enrichment, and an exponential regression model is proposed for fast dissolve CO_2 retrieval. The model feasibility is verified by performing a comparison test under two different situations, i.e., measurement chamber CO_2 concentration is higher and lower than the dissolved CO_2 concentration. Both their R-square of fitting are better than 98.5% and the difference of the two individually measured PDMS membrane coefficient $\frac{D_G S_G A}{lV}$ is only 3.8%. The whole absorption spectrum of CO_2 can be obtained within 90 s and a detection sensitivity of 1.8 ppm has been achieved. A near coast in-situ measurement has been implemented over 24 h and provided regular fluctuation of dissolved CO_2 concentrations, which is due to the tide phenomenon. Future efforts will be made to improve the corrosion resistance ability by using titanium alloy stainless steel as the pressured chamber material and to improve the gas separation efficiency by increasing the membrane surface area. Therefore, the developed senor could act as a promising tool to achieve high-precision detection of dissolved gas in seawater and then support the investigation on the ocean, such as vertical dissolved CO_2 profile as deep as 4500 m and long-term dissolved CO_2 monitoring under deep-sea extreme environment window, e.g., hydrothermal and cold spring.

Author Contributions: Conceptualization, M.H. and B.C.; methodology, M.H. and R.K.; software, M.H. and L.Y.; validation, M.H. and C.Y.; formal analysis, X.C.; investigation, X.C.; resources, C.Y.; writing—original draft preparation, M.H.; writing—review and editing, R.K. All authors have read and agreed to the published version of the manuscript.

Funding: This research was supported by the Scientific Instrument Developing Project of the Chinese Academy of Sciences (YJKYYQ20190037), the Second Comprehensive Scientific Investigation of the Qinghai-Tibet Plateau(2019QZKK020802), the National Key Research and Development Project (2019YFB2006003) and the National Natural Science Foundation of China (61805286).

Institutional Review Board Statement: Not applicable.

Informed Consent Statement: Not applicable.

Data Availability Statement: The data that support the plots within this paper are available from the corresponding author on request basis.

Conflicts of Interest: The authors declare no conflict of interest.

References

1. Tanhua, T.; Bates, N.R.; Körtzinger, A. The Marine Carbon Cycle and Ocean Carbon Inventories. In *Ocean Circulation and Climate A 21st Century Perspective*; Academic Press: Cambridge, MA, USA, 2013; pp. 787–815.
2. Dickens, G.R. Rethinking the global carbon cycle with a large, dynamic and microbially mediated gas hydrate capacitor. *Earth Planet. Sci. Lett.* **2003**, *213*, 169–183. [CrossRef]
3. Lechtenfeld, O.J.; Hertkorn, N.; Shen, Y.; Witt, M.; Benner, R. Marine sequestration of carbon in bacterial metabolites. *Nat. Commun.* **2015**, *6*, 6711. [CrossRef]
4. Wankel, S.D.; Germanovich, L.N.; Lilley, M.D.; Genc, G.; DiPerna, C.J.; Bradley, A.S.; Olson, E.J.; Girguis, P.R. Influence of subsurface biosphere on geochemical fluxes from diffuse hydrothermal fluids. *Nat. Geosci.* **2011**, *4*, 461–468. [CrossRef]
5. Nicholson, D.P.; Michel, A.P.M.; Wankel, S.D.; Manganini, K.; Sugrue, R.A.; Sandwith, Z.O.; Monk, S.A. Rapid Mapping of Dissolved Methane and Carbon Dioxide in Coastal Ecosystems Using the ChemYak Autonomous Surface Vehicle. *Environ. Sci. Technol.* **2018**, *52*, 13314–13324. [CrossRef] [PubMed]
6. Lohrenz, S.E.; Cai, W.J.; Chakraborty, S.; Huang, W.J.; Guo, X.; He, R.; Xue, Z.; Fennel, K.; Howden, S.; Tian, H. Satellite estimation of coastal pCO$_2$ and air-sea flux of carbon dioxide in the northern Gulf of Mexico. *Remote. Sens. Environ.* **2018**, *207*, 71–83. [CrossRef]
7. Michel, A.P.M.; Wankel, S.D.; Kapit, J.; Sandwith, Z.; Girguis, P.R. In situ carbon isotopic exploration of an active submarine volcano. *Deep Sea Res. Part II Top. Stud. Oceanogr.* **2018**, *150*, 57–66. [CrossRef]
8. Chua, E.J.; Savidge, W.; Short, R.T.; Cardenas-Valencia, A.M.; Fulweiler, R.W. A Review of the Emerging Field of Underwater Mass Spectrometry. *Front. Mar. Sci.* **2016**, *3*, 209. [CrossRef]
9. Wankel, S.D.; Joye, S.B.; Samarkin, V.A.; Shah, S.R.; Friederich, G.; Melas-Kyriazi, J.; Girguis, P.R. New constraints on methane fluxes and rates of anaerobic methane oxidation in a Gulf of Mexico brine pool via in situ mass spectrometry. *Deep Sea Res. Part II Top. Stud. Oceanogr.* **2010**, *57*, 2022–2029. [CrossRef]
10. Wang, Z.; Wang, Q.; Ching, J.Y.L.; Wu, J.C.-Y.; Zhang, G.; Ren, W. A portable low-power QEPAS-based CO$_2$ isotope sensor using a fiber-coupled interband cascade laser. *Sens. Actuators B Chem.* **2017**, *246*, 710–715. [CrossRef]
11. Wankel, S.D.; Huang, Y.W.; Gupta, M.; Provencal, R.; Leen, J.B.; Fahrland, A.; Vidoudez, C.; Girguis, P.R. Characterizing the distribution of methane sources and cycling in the deep sea via in situ stable isotope analysis. *Environ. Sci. Technol.* **2013**, *47*, 1478–1486. [CrossRef]

12. Liu, Z.; Zheng, C.; Zhang, T.; Li, Y.; Ren, Q.; Chen, C.; Ye, W.; Zhang, Y.; Wang, Y.; Tittel, F.K. Midinfrared Sensor System Based on Tunable Laser Absorption Spectroscopy for Dissolved Carbon Dioxide Analysis in the South China Sea: System-Level Integration and Deployment. *Anal. Chem.* **2020**, *92*, 8178–8185. [CrossRef]

13. Graziani, S.; Beaubien, S.E.; Bigi, S.; Lombardi, S. Spatial and temporal pCO2 marine monitoring near Panarea Island (Italy) using multiple low-cost GasPro sensors. *Environ. Sci. Technol.* **2014**, *48*, 12126–12133. [CrossRef] [PubMed]

14. Zhang, H.; Jin, W.; Hu, M.; Hu, M.; Liang, J.; Wang, Q. Investigation and Optimization of a Line-Locked Quartz Enhanced Spectrophone for Rapid Carbon Dioxide Measurement. *Sensors* **2021**, *21*, 5225. [CrossRef] [PubMed]

15. Xia, J.; Feng, C.; Zhu, F.; Ye, S.; Zhang, S.; Kolomenskii, A.; Wang, Q.; Dong, J.; Wang, Z.; Jin, W.; et al. A sensitive methane sensor of a ppt detection level using a mid-infrared interband cascade laser and a long-path multipass cell. *Sens. Actuators B Chem.* **2021**, *334*, 129641. [CrossRef]

16. Boulart, C.; Connelly, D.P.; Mowlem, M.C. Sensors and technologies for in situ dissolved methane measurements and their evaluation using Technology Readiness Levels. *TrAC Trends Anal. Chem.* **2010**, *29*, 186–195. [CrossRef]

17. Maity, A.; Maithani, S.; Pradhan, M. Cavity Ring-Down Spectroscopy: Recent Technological Advancements, Techniques, and Applications. *Anal. Chem.* **2021**, *93*, 388–416. [CrossRef] [PubMed]

18. Zhao, G.; Hausmaninger, T.; Ma, W.; Axner, O. Differential noise-immune cavity-enhanced optical heterodyne molecular spectroscopy for improvement of the detection sensitivity by reduction of drifts from background signals. *Opt. Express* **2017**, *25*, 29454–29471. [CrossRef]

19. Wu, H.; Stolarczyk, N.; Liu, Q.H.; Cheng, C.F.; Hua, T.P.; Sun, Y.R.; Hu, S.M. Comb-locked cavity ring-down spectroscopy with variable temperature. *Opt. Express* **2019**, *27*, 37559–37567. [CrossRef]

20. Hua, T.P.; Sun, Y.R.; Wang, J.; Liu, A.W.; Hu, S.M. Frequency metrology of molecules in the near-infrared by NICE-OHMS. *Opt. Express* **2019**, *27*, 6106–6115. [CrossRef]

21. Chen, Y.; Lehmann, K.K.; Kessler, J.; Sherwood Lollar, B.; Couloume, G.L.; Onstott, T.C. Measurement of the $^{13}C/^{12}C$ of atmospheric CH4 using near-infrared (NIR) cavity ring-down spectroscopy. *Anal. Chem.* **2013**, *85*, 11250–11257. [CrossRef]

22. Gupta, P.; Noone, D.; Galewsky, J.; Sweeney, C.; Vaughn, B.H. Demonstration of high-precision continuous measurements of water vapor isotopologues in laboratory and remote field deployments using wavelength-scanned cavity ring-down spectroscopy (WS-CRDS) technology. *Rapid Commun. Mass Spectrom.* **2009**, *23*, 2534–2542. [CrossRef] [PubMed]

23. Wang, Z.; Wang, Q.; Zhang, W.; Wei, H.; Li, Y.; Ren, W. Ultrasensitive photoacoustic detection in a high-finesse cavity with Pound-Drever-Hall locking. *Opt. Lett.* **2019**, *44*, 1924–1927. [CrossRef] [PubMed]

24. Ismail, A.F.; Kusworoa, D.T.; Mustafa, A.; Hasbullaha, H. Understanding the Solution-Diffusion Mechanism in Gas Separation Membrane for Engineering Students [C]. In Proceedings of the 2005 Regional Conference on Engineering Education, Fayetteville, AR, USA, 14–26 September 2005.

25. Lin, D.; Ding, Z.; Liu, L.; Ma, R. A Method to Obtain Gas-PDMS Membrane Interaction Parameters for UNIQUAC Model. *Chin. J. Chem. Eng.* **2013**, *21*, 485–493. [CrossRef]

26. Sysoev, A.A. A Mathematical Model for Kinetic Study of Analyte Permeation from Both Liquid and Gas Phases through Hollow Fiber Membranes into Vacuum. *Anal. Chem.* **2000**, *72*, 4221–4229. [CrossRef]

27. Bell, R.J.; Short, R.T.; Vanamerom, F.H.W.; Byrne, R.B. Calibration of an In Situ Membrane Inlet Mass Spectrometer for Measurements of Dissolved Gases and Volatile Organics in Seawater. *Environ. Sci. Technol.* **2007**, *41*, 8123–8128. [CrossRef]

28. Lehmann, K.K.; Berden, G.; Engeln, R. An Introduction to Cavity Ring-Down Spectroscopy. In *Cavity Ring-Down Spectroscopy*; John Wiley & Sons: Hoboken, NJ, USA, 2009; pp. 1–26.

29. Chen, J.-Y.; Liu, J.-G.; He, Y.-B.; Wang, L.; Gang, Q.; Xu, Z.-Y.; Yao, L.; Yuan, S.; Ruan, J.; He, J.-F.; et al. Study of CO2 spectroscopic parameters at high temperature near 2.0 µm. *Acta Phys. Sin.* **2013**, *62*, 224206.

30. Gordon, I.E.; Rothman, L.S.; Hill, C.; Kochanov, R.V.; Tan, Y.; Bernath, P.F.; Birk, M.; Boudon, V.; Campargue, A.; Chance, K.V.; et al. The HITRAN2016 molecular spectroscopic database. *J. Quant. Spectrosc. RA* **2017**, *203*, 3–69. [CrossRef]

Article

A Novel Airborne Molecular Contaminants Sensor Based on Sagnac Microfiber Structure

Guorui Zhou [1], Siheng Xiang [1], Hui You [1], Chunling Li [2], Longfei Niu [1], Yilan Jiang [1], Xinxiang Miao [1] and Xiufang Xie [2,*]

[1] Laser Fusion Research Center, China Academy of Engineering Physics, Mianyang 621900, China; zhougr@caep.cn (G.Z.); xiangsiheng1230@163.com (S.X.); yhworkyx@163.com (H.Y.); niulf@caep.cn (L.N.); jiangyilan1023@163.com (Y.J.); miaoxx@caep.cn (X.M.)

[2] Institute of Applied Electronics, China Academy of Engineering Physics, Mianyang 621900, China; odiaha@sina.com

* Correspondence: xie_xiu_fang@sina.com; Tel.: +86-0816-2491335

Abstract: The impact of airborne molecular contaminants (AMCs) on the lifetime of fused silica UV optics in high power lasers (HPLs) is a critical issue. In this work, we demonstrated the on-line monitoring method of AMCs concentration based on the Sagnac microfiber structure. In the experiment, a Sagnac microfiber loop with mesoporous silica coating was fabricated by the microheater brushing technique and dip coating. The physical absorption of AMCs in the mesoporous coating results in modification of the surrounding refractive index (RI). By monitoring the spectral shift in the wavelength domain, the proposed structure can operate as an AMCs concentration sensor. The sensitivity of the AMCs sensor can achieve 0.11 nm (mg/m^3). By evaluating the gas discharge characteristic of four different low volatilization greases in a coarse vacuum environment, we demonstrated the feasibility of the proposed sensors. The use of these sensors was shown to be very promising for meeting the requirements of detecting trace amounts of contaminants.

Keywords: airborne molecular contaminants (AMCs); Sagnac microfiber loop; high power lasers (HPLs)

Citation: Zhou, G.; Xiang, S.; You, H.; Li, C.; Niu, L.; Jiang, Y.; Miao, X.; Xie, X. A Novel Airborne Molecular Contaminants Sensor Based on Sagnac Microfiber Structure. *Sensors* **2022**, *22*, 1520. https://doi.org/10.3390/s22041520

Academic Editor: Jin Li

Received: 11 December 2021
Accepted: 1 February 2022
Published: 16 February 2022

Publisher's Note: MDPI stays neutral with regard to jurisdictional claims in published maps and institutional affiliations.

1. Introduction

Airborne molecular contaminants (AMCs) from material outgassing and shedding continue to be a challenging problem for optics components in high power lasers (HPLs). In order to avoid breakdown of air near the focal region on propagation of intense laser radiation, the optics components used as spatial filters/optical replays were mounted on in a vacuum environment. Compared to the atmosphere, AMCs are more easily produced by slow outgassing of material present in a vacuum environment, such as vacuum pump oil, cables, residual organic compounds of optics and mechanical component sub-surface. The AMCs result in the reduction in the laser-induced damage threshold (LIDT) of optics components [1–4] and the increase in reflectivity of the sol-gel porous silica antireflection coating [5]. Thus, online monitoring of the concentration of AMCs in a vacuum environment is a critical issue for maintaining the function of the optics components under "Safety Red line". Based on our research findings, the measured AMCs concentration ranged from 0.253 mg/m^3 to 46.228 mg/m^3 by using the gas chromatograph-mass spectrometer (GC-MS) technique during the operation of HPLs. On the other hand, other detection methods for AMCs are limited in harsh scenarios, especially in a vacuum or intense laser radiation environment [6,7]. Therefore, it is different to provide a more appropriate way for in-situ measurement of trace amounts of AMC in HPLs.

From the viewpoint of the optical fiber sensors structural design, sensors with thinner diameter structure have successfully achieved higher sensitivity [8]. Compared with the standard optical fiber for optical communication systems, optical microfibers (OMFs) have diameters ranging from hundreds of nanometers to tens of micrometers. Light guided along such OMFs leaves a large fraction of guided field outside the OMFs as evanescent waves to penetrate the environment, which allows the direct interaction between light and its surroundings [9–11]. Due to its thin diameter, the OMFs provide other commendable features, such as outstanding mechanical flexibility, tight optical confinement and low optical loss [12]. There are several major types of OMFs sensors that have been developed recently, which include Mach–Zehnder OMFs interferometric sensors [13–17], OMFs Bragg/long period grating (FBG)-based sensors [18–20], OMFs coupler sensors [21–24], Sagnac OMFs interferometric sensors [25–29], and OMFs multimode interferometric sensors [30–32]. The various types of sensors, depending on their unique structures, possess different sensing performances, in terms of sensitivity and temperature stability. Due to its truly path-matched interference mode, the Sagnac interferometric sensor has shown its superior temperature stability and high sensitivity, which renders it suitable, particularly for AMCs sensing.

In this paper, we proposed and investigated a novel AMCs sensor structure consisting of a Sagnac OMFs loop and a mesoporous silica coating on the OMFs surface. The experimental verification based on the proposed configuration was provided, along with the theoretical analysis. The AMCs sensors based on the Sagnac OMFs loop were fabricated by the heating brushing technique and by dip coating after synthesizing the sol-gel solutions. Verifying the proposed AMCs sensors indicated that they would have high sensitivity. In addition, to achieve the in-situ monitoring of typical AMCs, such as dibutyl phthalate (DBP), a gas discharge characteristic qualitative evaluation of various greases in a vacuum environment was done to verify the technical feasibility of the AMCs sensors. Furthermore, the sensors provided supporting measures of optics components operation, which is suitable for potential application in biological and chemical detection. Compared with our previous work [33], this work has made progress on novel structure and theoretical analysis, as well as the measurement method and technical feasibility verification. Firstly, taking the advantage of energy transfer in the coupling region, we reused the region as the effective sensing region; the device with the Sagnac OMFs loop possesses a more compact structure than a system containing an OMFs coupler. Secondly, the Sagnac OMFs loop acted as a reflector that reflects two beams from the OMFs coupler and that enhances interaction between the evanescent field and the external environment when the two beams re-enter OMFs and interfere. The sensors based on the Sagnac OMFs loop can achieve higher sensitivity. Therefore, we have carried out theoretical analysis and formular derivation for the OMFs sensor based on Sagnac interference. Finally, during the experiment, we optimized the experiment setup to further eliminate the measurement errors and to enhance the accuracy of the sensor performance measurement. Simultaneously, owing to its unique structure, the Sagnac OMFs loop is characterized by more symbolic wavelength dips, better recognizability in the measurement process, and better usability. In addition, its technical feasibility has been carried out in this work.

2. Sensor Configuration and Principle

Figure 1 shows the schematic configuration of the proposed AMCs sensor. It consisted of a Sagnac OMFs loop and mesoporous silica coating on the OMFs surface; the Sagnac OMFs loop contained an OMFs coupler and a Sagnac loop. The method of fabricating the Sagnac OMFs loop has already been studied in a previous report [33]. As shown in Figure 1, light injected into port P_0 could be coupled and split into clockwise and counterclockwise beams by the OMFs coupler, which propagates along the Sagnac loop; the interference was given by the recombination of the counter-propagating beams at the coupler. The principle of the sensor is that the beating between the lowest order even supermode and odd supermode provides energy transfer between two OMFs; wavelength dips in spectral region can be obtained as a result of the optical path difference between the two supermodes accumulated along the coupling region [34].

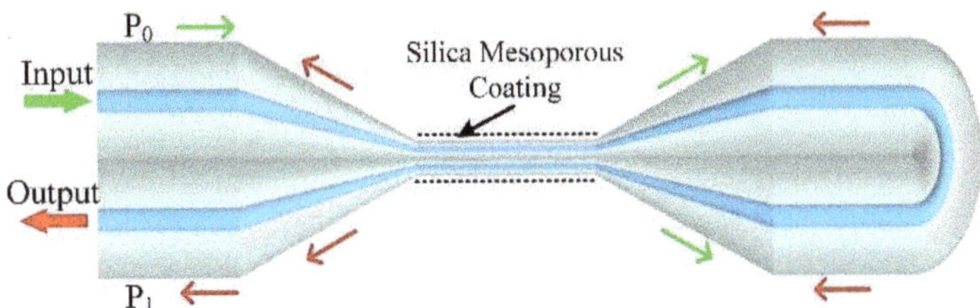

Figure 1. Schematic diagram of the AMCs sensor that was fabricated by employing the microheater brushing technique and dip coating.

In contrast, weakly fused OMFs couplers own a stronger dependency on ambient RIs than the strongly fused ones and are more suitable to work as sensing units [35]. Therefore, only weakly fused OMFs couplers were studied and fabricated in this thesis. Both OMFs roughly maintained original cross-section geometry throughout the entire fabrication process due to the larger aspect ratio, defined as the ratio between the total width and the height of the cross section in the uniform region (generally, larger than 1.8) [36]. Thus, it is reasonable as an acceptable simplification that two identical OMFs with a uniform shape and RI profile are fused together. Due to weakly fused OMFs in the uniform region, weak coupling theory was employed for the Sagnac OMFs loop sensing structure. The coupling coefficient C of the whole coupling region can be approximately denoted as [37]:

$$C(\lambda) = \frac{\pi\sqrt{n_1^2 - n_2^2}}{2an_1} e^{-2.3026(A+B\tau+C\tau^2)} \tag{1}$$

$$A = a_1 + a_2 V + a_3 V^2 \quad a_1 = 2.2926 \quad a_2 = -1.591 \quad a_3 = -0.1668$$
$$B = b_1 + b_2 V + b_3 V^2 \quad b_1 = -0.3374 \quad b_2 = 0.5321 \quad b_3 = -0.0066$$
$$C = c_1 + c_2 V + c_3 V^2 \quad c_1 = -0.0076 \quad c_2 = -0.0028 \quad c_3 = 0.0004$$

where λ is the wavelength of the incident light, n_1 and n_2 refer to the RIs of Sagnac OMFs loop and the surrounding medium, a is the radius of OMF in the waist region, $\tau = d/a$ is the aspect ratio of the cross section of the OMFs coupler, d is the distance between the axis of the fused OMFs and V is the normalized frequency where $V = [(2\pi a)/\lambda]\cdot(n_1^2 - n_2^2)^{1/2}$. The coefficients $a_1, a_2 \ldots c$ are functions of the shape of the OMFs couplers. One can approximate the modal field distribution of the fundamental mode by the Gaussian-exponential-Hankel function, whose parameters are determined by using a variable technique and the coupled-mode theory [37]. Based on these, we have found that, even for a step-index OMFs coupler, an empirical relation between the coupling coefficient C and V can again be described; the values of the coefficients for the OMFs coupler are given. When incident light of power enters the P_0 input port, the power at the P_1 output port can be expressed as [38]:

$$P_1 = \frac{1}{2}\left[1 + cos^2(2CL)\right] \tag{2}$$

where L is the coupling length of the Sagnac OMFs loop. According to Equation (1), the coupling coefficient C is related to the RI of the surrounding medium, the incident wavelength λ and the coupler radial size $2a$. Considering effective RI modification of mesoporous coating after the physical absorption of AMCs, the wavelength of the dip and intensity in the output interference spectrum varied as AMCs concentration in external environments changed when the silica mesoporous coating was used as the surrounding medium of the Sagnac OMFs loop. Therefore, the measurement of AMCs concentration could be realized by detecting the wavelength shift of the dip and variation in the spectral domain. According to Equations (1) and (2), the sensitivity of the wavelength shift of the dip to the RI change of the surrounding medium can be derived as

$$\frac{\partial \lambda}{\partial n_2} = -\frac{\partial(2CL)/\partial n_2}{\partial(2CL)/\partial \lambda} = \frac{4\pi^2 a^2 n_2}{V^2}\left(\frac{1}{DV\lambda} - 1\right) \tag{3}$$

$$D = a_2 + 2a_3 V + b_2 + 2b_3 V + c_2 + 2c_3 V$$

Here, V is the normalized frequency, a is the radius of OMF in the waist region, and there is a linear relationship between D and V. For the Sagnac OMFs loop structure, $V > 1$. Therefore, it is easy to obtain $\frac{\partial \lambda}{\partial n_2} < 0$, which indicates that the wavelength of the dips blueshifts as the RI of the surrounding medium n_2 increases. Besides, according to the results of numerical simulation for single mode fiber [12], there are various modes existing in the coupling region when $V > 1$ and the diameter of OMFs is greater than 800 nm, even appearing in several high-order modes (e.g., HE31). After preliminary calculation, when the diameter of the waist region of Sagnac OMFs loop was 3.0 μm and the incident wavelength was 1550 nm, there were more than 10 modes in the coupling region.

3. Experiment

3.1. Silica Sensitive Coating Solution Preparation

Here, we employed a facile one-step solution-based synthetic route to produce silica sensitive coating material using controlled hydrolysis of tetraethyl orthosilicate (TEOS) via the Stober method. Firstly, base-catalyzed silica solution was fabricated by stirring the mixture consisting of EtOH, TEOS, H_2O and $NH_3 \cdot H_2O$ for 2 h at 30° and aged at 25° for 7 days; the molar ratio of EtOH, TEOS, H_2O and $NH_3 \cdot H_2O$ was 1:3.25:37.6:0.17. After refluxing for 24 h to remove ammonia and filtering through a 0.22 μm PVDF filter, the final concentration of SiO_2 in the solution was 3% by weight.

Figure 2 illustrates an SEM image of fabricated SiO_2 nanoparticles utilizing the Stober method. As shown in the SEM image, the shapes of silica nanoparticles with a diameter of 20.0 nm were not very regular, and the mesoporous coating formed on the glass slide substrate by dense packing and or a van der Waals attraction. In addition, porosity increased due to abundant voids of coatings [39].

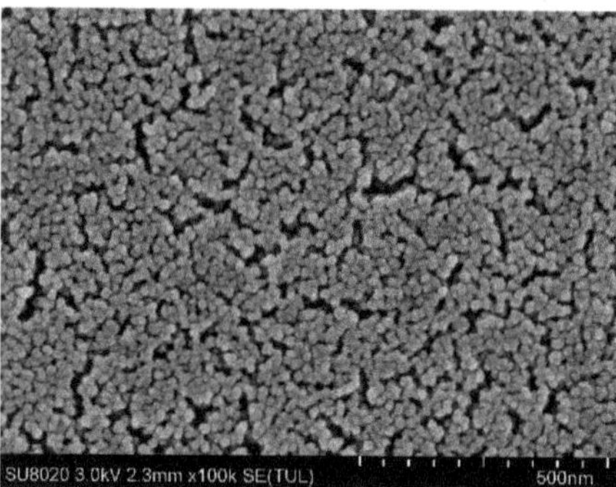

Figure 2. SEM image of fabricated SiO_2 nano-particles utilizing Stober method.

3.2. Fabrication of Sagnac OMFs Loop Sensing Unit and Experimental Setup

To fabricate the Sagnac OMFs loop, a piece of standard telecom single-mode fibers (SMF28, Corning) was firstly bent, and the two free ends were twisted together carefully to form a Sagnac loop. In order to ensure that the fibers remained in contact commendably and to prevent fibers fracturing during fabrication, the distance between two adjacent fiber knots should be kept from 4 mm to 6 mm. The twisted region was entirely fused and tapered using the microheater brushing technology. The tapering process was executed on the homemade taper drawing system, which consists of three translation stages driven by linear motors with 150 nm of positioning accuracy. In order to obtain the Sagnac OMFs loop with the required shape and diameter in the waist region, the fabrication process of a low-loss Sagnac OMFs loop was accurately controlled by a computer program.

Considering their outstanding compatibility and stability, aluminum alloy and polydimethylsiloxane (PDMS) were employed to package the proposed Sagnac OMFs loop in the aluminum alloy holder. Based on the dip coating method, the packaged Sagnac OMFs loop was dipped into prepared SiO_2 solution mentioned above for at least 1 min, and permeated totally. The silica solution concentration and pulling rate can control the thickness of the SiO_2 coating on the surface of the Sagnac OMFs loop. In the coating fabrication process, the fabrication setup moved upwards at the pulling speed of 200 mm/min; in order to enhance the mechanical properties of the coating, the proposed Sagnac OMFs loop with mesoporous coating was exposed in hexamethyldisilazane (HDMS) vapor in a clean room (ISO 5) by converting the remaining hydroxyl groups to trimetrylsiloxy groups. The geometry and surface morphology of the proposed sensing unit was examined under a scanning electron microscope (SEM). The SEM image of a short section of the sensing unit was shown in Figure 3a. Both fused OMFs that roughly maintained their original geometry further confirmed the weakly fused status of the fabricated Sagnac loop. Each OMFs possessed a diameter of ~2.87 μm and showed very outstanding diameter uniformity in the coupling region. The length of the waist was about 7 mm. Due to the partial overlapping of two OMFs, it could not see the complete shape of the other OMFs. Figure 3b shows a mesoporous silica coating on the surface of the fiber. An SEM image of the cross section of the proposed Sagnac OMFs loop sensing unit is illustrated in Figure 3c. Figure 3c shows the fabricated mesoporous silica coating on the surface of the Sagnac OMFs loop; the thickness of the coating was ~88.2 nm.

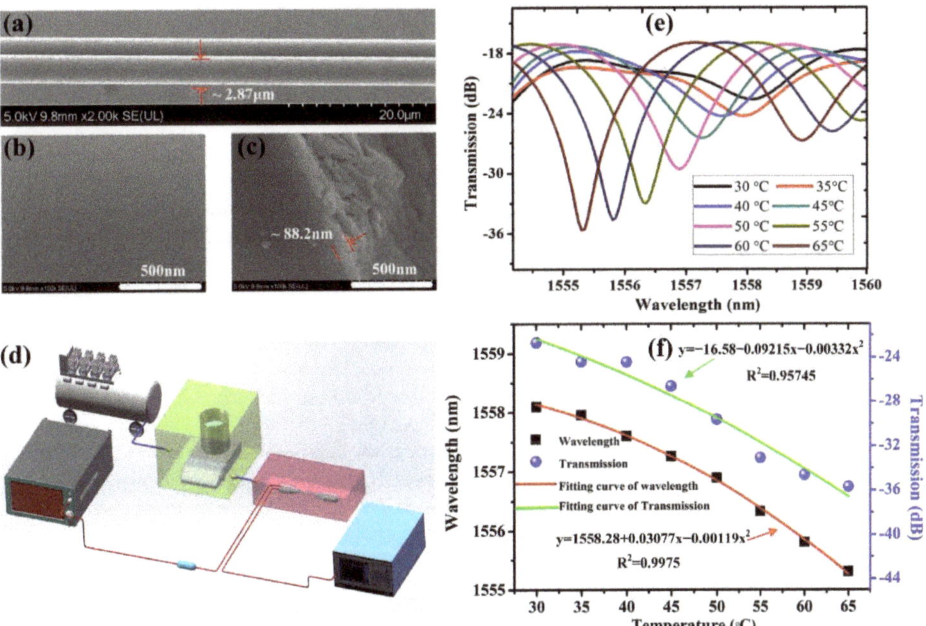

Figure 3. (**a**) shows a typical SEM image of uniform region of OMFs coupler; (**b**) shows microtopography of silica mesoporous coating on OMFs coupler; (**c**) shows a typical SEM image of the cross section of the fabricated sensing unit; (**d**) the schematic of detection equipment; (**e**) transmission spectra of the proposed sensor at different Ts; (**f**) response of the wavelength and intensity of the dip under different Ts.

Due to poor mechanical stability, aluminum alloy holders with a micro-channel for washing the sensing unit were machined to fix the sensing unit. The holders did not produce any contaminants to eliminate measurement error, especially after ultrasonic cleaning. The experimental setup for the characterization of the AMCs sensor is described in Figure 3d. Incident light emitted from a semiconductor laser diode (SLD, Thorlabs Inc., Newton, NJ, USA, S5FC1550P-A2) with a center wavelength of 1550 nm was injected into the input port P_0; the output light from P_1 was detected by an optical spectrum analyzer with a resolution of 0.02 nm (AQ6370, Yokogawa Co. Ltd., Tokyo, Japan). The sensing units were intentionally contaminated by outgassing DBP; the aim of this experiment was to simulate an optics contamination phenomenon in order to determine contamination kinetics and the proposed AMCs sensors sensitivity. DBP was used as a model AMC for the present experiments. Owing to extensive use as plasticizers, DBP and its less volatile homologue dioctalphthalate (DOP) are very common during the building of HPLs. Moreover, DBP was also used in this experiment because of its convenient transport properties and easy diffusion in a common environment. The material was outgassed at a heating temperature that ranged from 30 °C to 70 °C; the clean compressed air (ISO 3) charged with contaminants was directed towards the proposed sensors at a flow rate of 2 L/min in order to help condensation of organics. A weighing bottle filled with DBP was heated by a magnetic stirrer with precise control of temperature (IKA RET/T). The magnetic stirrer and the proposed sensors were put in two chambers with a gas inlet and a gas outlet, respectively. We delivered the clean compressed air into the heating chamber; the mixture of the clean air and AMCs led into the measurement chamber further affected the properties of the sensors. In the measurement process, at a heating temperature, we found that the volatilization rate of DBP was basically constant (except for the first two minutes). Thus, it is reasonable to believe that the concentration of AMCs in the mixture was basically constant at the heating temperature. When a certain concentration of mixed gas gradually filled the measurement chamber, the excess gas would also be discharged through the outlet. Finally, at the heating temperature, a constant AMCs concentration would be found in the measurement chamber. Therefore, it can be seen that the heating temperature corresponded to a specific AMCs concentration.

It is important to investigate temperature characteristics of the proposed sensor in its application environment, in which they exhibit inaccuracies under special conditions, such as high-humidity and an environment with strong electromagnetic interference. The proposed sensor was placed in a thermostat with precise control of temperature and a commercial temperature sensor (Rotronic HC2A-S). As shown in Figure 3e, as the T increased from 30 °C to 65 °C, the wavelength dip exhibited a redshift and a decrease in intensity due to the combined effects of the thermo-optics and the thermal expansion in the tapered waist region. Here, since the diameter of OMFs is very thin, the fiber core can be neglected [40]. The change of fiber cladding owing to thermo-optic effect in the waist region plays a key role. Thus, an increase in T mainly results in an increase in the RI of the fiber cladding [41]. Figure 3f illustrates the response of the wavelength and intensity of the dip as ambient temperature T increased. It suggests polynomial fitting with an R-square of 0.9975 and 0.95745 for wavelength and intensity of the dip, respectively. Simultaneously, the fitting coefficients of wavelength and intensity of the dip were 0.03077, −0.00119 and −0.09215, −0.00332, respectively. When the ambient temperature increased from 30 °C to 65 °C, the wavelength dip blueshifted by 2.79 nm. Moreover, the transmission loss of the dip increased by 13.02 dB. The wavelength shift and intensity of the dip possess a complex polynomial relationship with T change; T sensitivities of approximately −79.7 pm/°C and −0.372 dB/°C, respectively, were achieved. When ambient temperature varied, the change of the phase difference induced by the combined effects of the thermo-optics and the thermal expansion in the tapered waist region was dominant in the process. In addition, the phase difference depended on birefringent coefficient B and the length of the twist area L. The free spectral range (FSR) of the transmission spectrum would become narrower and narrower, as the number of twist turns increased. In this situation, considering that the birefringence coefficient B mainly depends on the shape of the twist area's cross-section and the temperature almost has no influence on B, $\partial B/\partial T$ could be regarded as zero. As a result, we were able to eliminate or solve the cross-sensitivity between AMCs and temperature by adjusting the number of twist turns in the proposed sensors preparation; theory derivation and experimental demonstration can be seen in detail in Reference [42].

3.3. Results and Discussion

In order to accurately obtain the concentration of DBP at different heating temperature, weight difference before and after heating was measured with micro-analytical balance (Mettler Toledo XP56) five times, to obtain the average value of weight difference. Considering the flow rate of the loading gas, the concentration of AMCs could be obtained. In order to verify the response time of the proposed sensors, the duration time of heating DBP was set to 20 min, and then the gas path and chambers were flushed by the clean air to avoid the impact of the condensation of AMCs on experimental results. The stability/repeatability of the proposed sensor mainly depended on the profile of OMFs and the influence of the background environmental cleanliness. In the fabrication process of Sagnac Loop OMFs, the evolution of the transmission spectrum of the tapered samples has been adopted to obtain required samples accurately. Meanwhile, we carried out high-spraying and ultrasonic cleaning of the chamber and its affiliated pipelines to avoid the influence of the background environmental contaminants, as mentioned above. Figure 4 illustrates transmission spectra evolution of the proposed sensors with 3.0 μm diameter at 30 °C, 50 °C and 70 °C. At heating temperatures of 30 °C, 50 °C and 70 °C, the wavelength dips shift approached shorter wavelength as the AMCs concentration increased. Simultaneously, it is noted that the higher the heating temperature, the larger the wavelength dip shift. Figure 4 also shows that the theoretical analysis has a good match with the experimental results. The experimental results shown in Figure 4 can be explained by RI variation of the silica sensitive coating and the polarization state of the input light in the waist region. The adsorption number of AMCs in the mesoporous coating increased with the increase in the concentration of DBP inside the measurement chamber, which resulted in the variation of coupling coefficients of the Sagnac loop. This caused the wavelength of the dip blueshifts on transmission spectrum. One factor that must be accounted for is that the polarization state of the input light changed as the adsorption number of AMCs increased in the sensing process. This would affect the spectral response; the overall spectrum was modified such that some peaks were considerably attenuated, even when distorted at 70 °C.

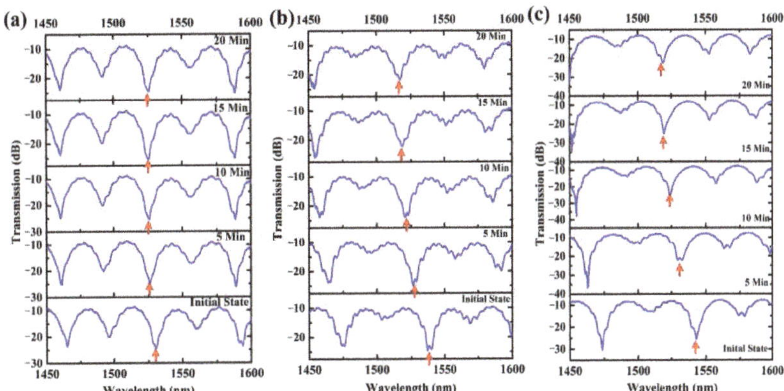

Figure 4. Transmission spectra evolution of the proposed sensor with 3.0 μm diameter at (**a**) 30 °C, (**b**) 50 °C and (**c**) 70 °C, the duration time of heating was 20 min in the experiment. The red arrows in the spectral responses represent corresponding dip wavelength shifts as elapsed time increases.

Figure 5a shows the wavelength dips shift throughout the experiment. The wavelength dip shifts of 5.4 nm, 17.9 nm and 24.8 nm correspond to the heating temperature of 30 °C, 50 °C and 70 °C, respectively. When the AMCs concentration was increasing, it appeared as a sharp decline and then became steady quickly, revealing good stability. As shown in Figure 5a, we could observe that the proposed sensors showed almost the same response time (about 14 min) at different heating temperatures, due to the same thickness of the mesoporous silica coating of the sensors. At the same concentration of AMCs, for the sensitive coating with the same thickness, it reached equilibrium between adsorption and desorption of AMCs in the same amount of time. In order to further demonstrate the experimental results, we prepared the AMCs sensors with 2.5 μm diameter at the same pulling speed. The results indicated that the response time of the proposed sensors with 2.5 μm diameter was reduced to 7.5 min. The thinner sensitive coating of the proposed sensors with 2.5 μm diameter is a major reason why the sensors own the shorter response time; compared with the thicker coating, the thinner coating more easily reached equilibrium between adsorption and desorption of AMCs under the same condition.

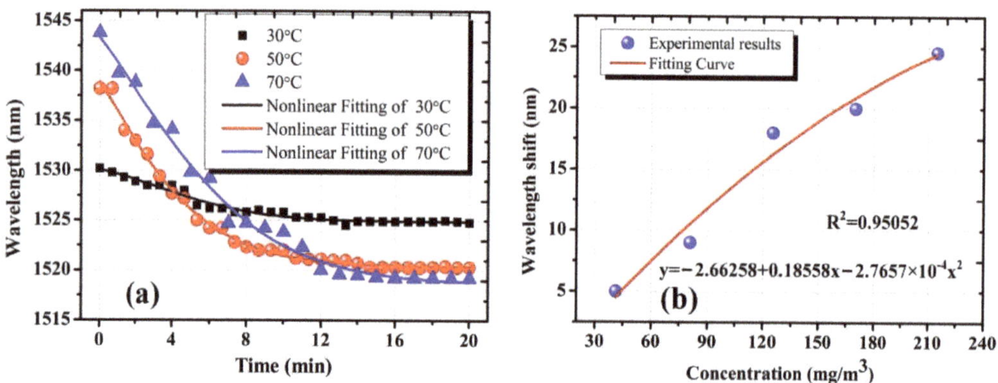

Figure 5. (a) The wavelength dips shift for the proposed sensors with 3.0 μm diameter as measurement time elapsed at 30 °C, 50 °C and 70 °C; (b) the relationship between wavelength shift and AMCs concentration for the as-fabrication sensors with 3.0 μm diameter.

Figure 5b plots the measured wavelength dip change as a function of the concentration of AMCs at different heating temperatures. In the experiment, the heating temperature varied from 30 °C to 70 °C at intervals of 10 °C. In order to achieve equilibrium between adsorption and desorption of AMCs, the duration time of heating was set at 20 min and the spectral response at each temperature was recorded. It showed a polynomial fitting with an R-square of 0.95052; the fitting coefficients were 0.18558 and -2.7657×10^{-4}. The wavelength dip shifted from 5.4 nm to 24.8 nm as the concentration increased from 0 to 213 mg/m^3. The average sensitivity was ~0.11 nm/(mg/m^3). Compared with the previous microfiber structure, the mesoporous silica coating allows for a stronger evanescent-field interaction between AMCs and the sensing waveguide, due to its high porosity and limited thickness.

3.4. The Technical Characteristic of the Proposed Sensors

The technical characteristics of the proposed sensors are closely interrelated to adsorption of the mesoporous sensitive coating, whose porosity directly determines the number of AMCs adsorbed in the sensing process. Consequently, it is reasonable to assume that a single OMF sensor with the same parameters possesses the same technical characteristics as the Sagnac loop sensor discussed here. In order to obtain characteristics of the Sagnac loop sensor, we prepared the single OMFs sensors with the same parameters. In the experiment, a piece of standard telecom single-mode fibers (SMF, Corning) was fixed on the homemade taper drawing system and tapered into OMF with a 3.0 μm diameter. Although we could accurately control the diameter and shape of OMFs through the taper drawing system, the prepared samples were be placed under the optical microscope to check whether its diameter met the requirement, so as to reduce the error induced by different diameters. The coating preparation method and parameters of single OMFs were completely consistent with those of Sagnac Loop sensors.

The physical adsorption of AMCs in the mesoporous coating of OMFs resulted in modification of the surrounding of RI; the additional loss was produced due to the absorption of the evanescent field. In the experiment, the additional loss caused by the adherence of AMCs was recorded in real time. A DFB laser at 1310 nm center wavelength was divided into two beams by a light coupler, one of which was used as the signal light and the other as the reference light to eliminate the error caused by laser power fluctuation. Furthermore, in order to improve measurement accuracy, lock-in amplification technology was adopted in the process of optical signal detection and processing. The photodetector, A/D converter and DSP mode-locked amplifier were integrated into a complete measurement system with a dynamic measurement range of about 25 dB; our own software was developed to process and store the monitoring data. We employed the experimental setup shown in Figure 3d to evaluate technical characteristics of the OMFs sensors. The heating temperature was gradually increased to reach the maximum measurement concentration of the OMFs sensors. When the heating temperature of AMCs was 98 °C, the response of the OMFs sensor was shown in Figure 6. Once the additional loss achieved the maximum value of the measurement system, heating stopped immediately and the AMCs concentration of 438 mg/m^3 was obtained through the weighing method mentioned above. One can see that the sensor possessed a maximum concentration of 438 mg/m^3 and a response time of 17 min, which was slightly larger than the experimental results of the Sagnac Loop sensors discussed above. Thus, it can be considered that the fabricated Sagnac Loop sensors also possessed a maximum concentration of 438 mg/m^3. However, we employed the heating method to realize the volatilization of AMCs; the minimum concentration of the proposed sensors could not be achieved due to control accuracy and condensation of AMCs in the gas path. We could obtain the sensing system resolution of 0.18 mg/m^3, resulting from an average sensitivity of 0.11 nm/(mg/m^3) and the OSA with the resolution of 0.02 nm used in the experiment.

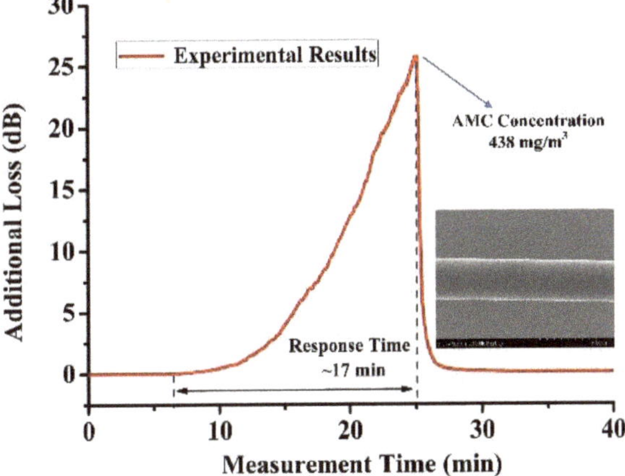

Figure 6. Response of the OMFs sensors at heating temperature of 98 °C, the inset shows the SEM image of the OMFs sensor with ~3.0 μm diameter.

As mentioned above, compared with other types of sensors, the Sagnac Loop sensor possesses more symbolic wavelength dips and better recognizability, especially for the sensor with a diameter of 3.0 μm. According to [43], the interval between the two adjacent characteristic wavelength dips was 70 nm, which was larger than the shift of wavelength dips induced by the maximum concentration of AMCs (about 48 nm). Therefore, based on our actual detection requirements, the concentration of AMCs in HPLs was less than 200 mg/m^3; the dynamic range of the sensing system in the spectrum domain could fully meet the detection requirements of maximum concentration.

3.5. The Feasibility of the Proposed Sensors

In order to validate the feasibility of the proposed sensors, the proposed sensor with 3.0 μm diameter was employed to qualitatively evaluate the gas discharge characteristic of four different low volatilization greases in a coarse vacuum environment. In the experiment, the measured grease samples were put into a vacuum chamber, connected to a dry vacuum pump (Agilent TriScroll 300) with high pumping speed. Before the experiment, in order to avoid the influence of the background environmental cleanliness, we carried out high-spraying and ultrasonic cleaning of the chamber and its affiliated pipelines. First, we utilized the vacuum pump to suction out the air; the negative pressure in the vacuum chamber was maintained at ~20 Pa. The negative gauge pressure was measured using a vacuum gauge (Televac CC-10) in the vacuum chamber. Then, via a switch air hole, air was let into the vacuum chamber to achieve the fine-tuning of vacuum degree; the spectra were recorded simultaneously. The duration time of the pumping process was 20 min. Transmission spectra comparison of grease samples 00#, 02#, 82# and 83# before and after evaluation are shown in Figure 7. We found different wavelength dip shifts for grease samples 00#, 02#, 82# and 83#, as well as a certain degree of spectra distortion due to transmission fiber deformation in the vacuum environment.

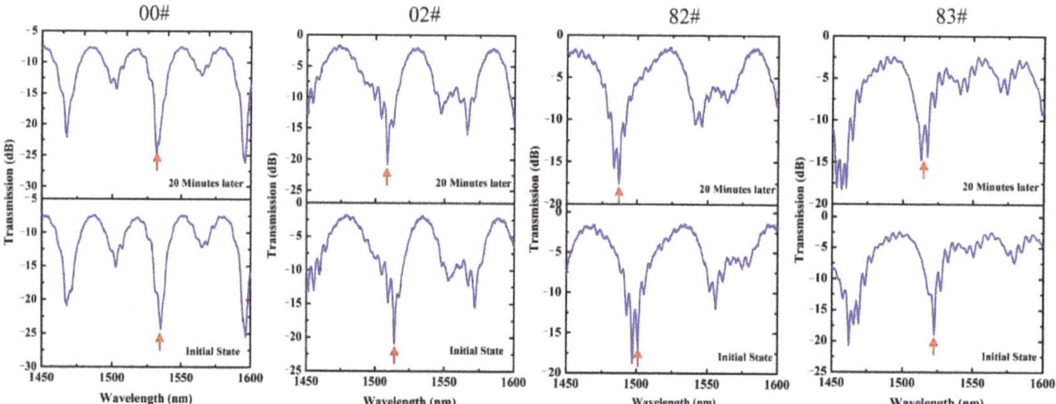

Figure 7. Transmission spectra evolution of grease sample 00#, 02#, 82# and 83# at coarse vacuum of ~20 Pa, and the diameter of the sensors used in the measurement process is 3.0 μm.

Figure 8 illustrates the wavelength dips shift of grease samples 00#, 02#, 82# and 83# at a coarse vacuum of ~20 Pa. For grease sample 00#, the wavelength dip shift was about 4 nm, less than the dip shift of other grease samples. The wavelength dip shift of sample 82# was the largest, about 12 nm. The dips shift of samples 02# and 83# were 5.4 nm and 7.5 nm, respectively, between the dips shift of samples 00# and 82#. From the experimental results, it can be clearly seen that grease sample 00#, with less volatilization, was more suitable for the vacuum environment of HPLs, while grease sample 82# was the opposite. Moreover, the results demonstrate the feasibility of the proposed sensors.

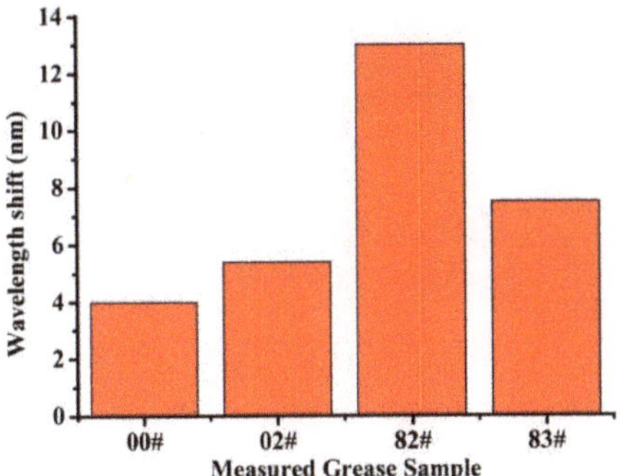

Figure 8. The wavelength dips shift of grease sample 00#, 02#, 82# and 83# at coarse vacuum of ~20 Pa.

4. Conclusions

An AMCs sensor based on the Sagnac microfiber structure for molecular contaminants concentration on-line measurement is presented. The device is fabricated by the microheater brushing technique and dip coating. For the DBP measurement as a model AMCs, the average sensitivity of the sensors was 0.11 nm/(mg/m^3). Gas discharge characteristic qualitative evaluation of several greases in the vacuum environment was done to verify the technical feasibility of the AMCs sensors. The sensing performance of the proposed sensor can be further improved through the parameter optimization of the structure. The sensor is inexpensive and easy to fabricate; thus, it has shown great potential for in-situ measurement of AMCs in HPLs.

Author Contributions: G.Z. and X.X. conceived and designed the experiments. G.Z., S.X. and H.Y. performed the experiments. G.Z. and S.X. performed the theoretical analysis. S.X., G.Z. and C.L. analyzed the data. G.Z. prepared the initial version of the paper. L.N., Y.J. and X.M. revised the paper. Everyone contributed to the results discussion, evaluation and edited the paper. All authors have read and agreed to the published version of the manuscript.

Funding: This research was funded by National Natural Science Foundation of China grant number [12174355, 61605186, 61705205] and Laser Fusion Research Center Funds for Young Talents grant number [RCFPD1-2017-7].

Institutional Review Board Statement: Not applicable.

Informed Consent Statement: Not applicable.

Data Availability Statement: The datasets generated during and/or analyzed during the current study are available from the corresponding author on reasonable request.

Acknowledgments: The work is supported by National Natural Science Foundation of China (Grant No. 12174355, 61605186, 61705205), Laser Fusion Research Center Funds for Young Talents (Grant No. RCFPD1-2017-7).

Conflicts of Interest: The authors declare no conflict of interest.

References

1. Bien-Aimé, K.; Belin, C.; Gallais, L.; Grua, P.; Tovena-Pecault, I. Impact of storage induced outgassing organic contamination on laser induced damage of silica optics at 351 nm. *Opt. Express* **2009**, *17*, 18703–18713. [CrossRef]
2. Yun, C.; Zhao, Y.; Hua, Y.; He, H.; Shao, J. Impact of organic contamination on laser-induced damage threshold of high reflectance coatings in vacuum. *Appl. Surf. Sci.* **2008**, *254*, 5990–5993.
3. Ling, X.; Zhao, Y.; Shao, J.; Fan, Z. Effect of two organic contamination modes on laser-induced damage of high reflective films in vacuum. *Thin Solid Film.* **2010**, *519*, 296–300. [CrossRef]
4. Bien-Aimé, K.; Néauport, J.; Tovena-Pecault, I.; Fargin, E.; Labrugère, C.; Belin, C.; Couzi, M. Laser induced damage of fused silica polished optics due to a droplet forming organic contaminant. *Appl. Opt.* **2009**, *48*, 2228–2235. [CrossRef]
5. Pareek, R.; Kumbhare, M.N. Effect of oil vapor contamination on the performance of porous silica sol-gel antireflection-coated optics in vacuum spatial filters of high-power neodymium glass laser. *Opt. Eng.* **2008**, *47*, 023801. [CrossRef]

6. Guo, L.; Shen, J. Dependence of the forward light scattering on the refractive index of particles. *Opt. Laser Technol.* **2018**, *101*, 232–241. [CrossRef]

7. Chen, A.; Wang, S.; Liu, Y.; Yan, S.; Deng, T. Light scattering intensity field imaging sensor for in-situ aerosol analysis. *ACS Sens.* **2020**, *5*, 2061–2066. [CrossRef]

8. Xiang, S.; You, H.; Miao, X.; Niu, L.; Yao, C.; Jiang, Y.; Zhou, G. An Ultra-Sensitive Multi-Functional Optical Micro/Nanofiber Based on Stretchable Encapsulation. *Sensors* **2021**, *21*, 7437. [CrossRef]

9. Lou, J.; Tong, L.; Ye, Z. Modeling of silica nanowires for optical sensing. *Opt. Express* **2005**, *13*, 2135–2140. [CrossRef]

10. Sun, Q.; Sun, X.; Jia, W.; Xu, Z.; Luo, H.; Liu, D.; Zhang, L. Graphene-Assisted Microfiber for Optical-Power-Based Temperature Sensor. *IEEE Photonics Technol. Lett.* **2015**, *28*, 383–386.

11. Li, X.; Ding, H. A Stable Evanescent Field-Based Microfiber Knot Resonator Refractive Index Sensor. *IEEE Photonics Technol. Lett.* **2014**, *26*, 1625–1628. [CrossRef]

12. Tong, L.; Lou, J.; Mazur, E. Single-mode guiding properties of subwavelength-diameter silica and silicon wire waveguides. *Opt. Express* **2003**, *12*, 1025–1035. [CrossRef] [PubMed]

13. Zhu, X.; Geng, J.; Sun, D.; Ji, Y.; Zhang, G. A novel Optical Fiber Sensor Based on Microfiber Mach-Zehnder Interferometer with Two Spindle-Shaped Structures. *IEEE Photonics J.* **2021**, *13*, 1. [CrossRef]

14. Wang, Y.; Zhang, H.; Cui, Y.; Duan, S.; Liu, B. A complementary-DNA-enhanced fiber-optic sensor based on microfiber-assisted Mach-Zehnder interferometry for biocompatible pH sensing. *Sens. Actuators B Chem.* **2021**, *332*, 129516. [CrossRef]

15. Ahsani, V.; Ahmed, F.; Jun, M.; Bradley, C. Tapered Fiber-Optic Mach-Zehnder Interferometer for Ultra-High Sensitivity Measurement of Refractive Index. *Sensors* **2019**, *19*, 1652. [CrossRef] [PubMed]

16. Zhang, H.; Cong, B.; Zhang, F.; Qi, Y.; Hu, T. Simultaneous measurement of refractive index and temperature by Mach-Zender cascaded with FBG sensor base on milti-core microfiber. *Opt. Commun.* **2021**, *493*, 126985. [CrossRef]

17. Ran, Z.; He, X.; Rao, Y.; Sun, D.; Qin, X.; Zeng, D.; Chu, W.; Li, X.; Wei, Y. Fiber-optic microstructure sensors: A review. *Photonic Sens.* **2021**, *11*, 227–261. [CrossRef]

18. Ran, Y.; Long, J.; Xu, Z.; Hu, D.; Guan, B.O. Temperature monitorable refractometer of microfiber Bragg grating using a duet of harmonic resonances. *Opt. Lett.* **2019**, *44*, 3186–3189. [CrossRef]

19. Zhang, X.; Zou, X.; Luo, B.; Pan, W.; Peng, W. Optically functionalized microfiber Bragg grating for RH sensing. *Opt. Lett.* **2019**, *44*, 4646–4649. [CrossRef]

20. Ran, Y.; Hu, D.; Xu, Z.; Long, J.; Guan, B.O. Vertical-fluid-array induced optical microfiber long period grating (VIOLIN) refractometer. *J. Lightwave Technol.* **2020**, *38*, 2434–2440. [CrossRef]

21. Peng, Y.; Zhao, Y.; Hu, X.G.; Chen, M.Q. Humidity sensor based on unsymmetrical U-shaped twisted microfiber coupler with wide detection range. *Sens. Actuators B Chem.* **2019**, *290*, 406–413. [CrossRef]

22. Wei, F.; Liu, D.; Mallik, A.K.; Farrel, G.; Wu, Q.; Peng, G.D.; Semenova, Y. Magnetic Field Sensor Based on a Tri-Microfiber Coupler Ring in Magnetic Fluid and a Fiber Bragg Grating. *Sensors* **2019**, *19*, 5100. [CrossRef]

23. Zhang, L.; Wang, Y.; Wu, H.; Hou, M.; Wang, Y. A ZnO nanowire-based microfiber coupler for all-optical photodetection applications. *Nanoscale* **2019**, *11*, 8319–8326. [CrossRef]

24. Wang, J.; Li, X.; Ju, J.; Li, K. High-Sensitivity, Large Dynamic Range Refractive Index Measurement Using an Optical Microfiber Coupler. *Sensors* **2019**, *19*, 5078. [CrossRef]

25. Sun, L.P.; Yuan, Z.; Huang, T.; Sun, Z.; Guan, B.O. Ultrasensitive sensing in air based on Sagnac interferometer working at group birefringence turning point. *Opt. Express* **2019**, *27*, 29501–29509. [CrossRef]

26. Li, X.; Nguyen, L.V.; Zhao, Y.; Heike, E.-H.; Warren-Smith, S.C. High-sensitivity Sagnac-interferometer biosensor based on exposed core microstructured optical fiber. *Sens. Actuators B Chem.* **2018**, *269*, 103–109. [CrossRef]

27. Gao, S.; Sun, L.P.; Jie, L.; Jin, L.; Guan, B.O. High-sensitivity DNA biosensor based on microfiber Sagnac interferometer. *Opt. Express* **2017**, *25*, 13305–13313. [CrossRef]

28. Chowdhury, S.; Verma, S.; Gangopadhyay, T.K. A comparative study and experimental observations of optical fiber sagnac interferometric based strain sensor by using different fibers. *Opt. Fiber Technol.* **2019**, *48*, 283–288. [CrossRef]

29. Liu, T.; Zhang, H.; Xue, L.; Liu, B.; Liu, H.; Huang, B.; Sun, J.; Wang, D. Highly sensitive torsion sensor based on side-hole-fiber Sagnac interferometer. *IEEE Sens. J.* **2019**, *19*, 7378–7382. [CrossRef]

30. Chen, Y.; Han, Q.; Liu, T.; Lue, X. Self-temperature-compensative refractometer based on singlemode–multimode–singlemode fiber structure. *Sens. Actuators B Chem.* **2015**, *212*, 107–111. [CrossRef]

31. Liu, D.; Mallik, A.K.; Yuan, J.; Yu, C.; Farrel, G.; Semenova, Y.; Wu, Q. High sensitivity refractive index sensor based on a tapered small core single-mode fiber structure. *Opt. Lett.* **2015**, *40*, 4166–4169. [CrossRef]

32. Wang, X.; Wang, J.; Wang, S.S.; Liao, Y.P. Fiber-Optic Salinity Sensing With a Panda-Microfiber-Based Multimode Interferometer. *J. Lightwave Technol.* **2017**, *35*, 5086–5091. [CrossRef]

33. Zhou, G.; Niu, L.; Jiang, Y.; Liu, H.; Lv, H. Sensing of airborne molecular contaminants based on microfiber coupler with mesoporous silica coating. *Sens. Actuators A Phys.* **2019**, *287*, 1–7. [CrossRef]

34. Li, K.; Zhang, T.; Liu, G.; Zhang, N.; Zhang, M.; Wei, L. Ultrasensitive optical microfiber coupler based sensors operating near the turning point of effective group index difference. *Appl. Phys. Lett.* **2016**, *109*, 101101. [CrossRef]

35. Lamont, R.G.; Johnson, D.C.; Hill, K.O. Power transfer in fused biconical-taper single-mode fiber couplers: Dependence on external refractive index. *Appl. Opt.* **1985**, *24*, 327–332. [CrossRef]

36. Morishita, K.; Yamazaki, K. Wavelength and polarization dependences of fused fiber couplers. *J. Lightwave Technol.* **2011**, *29*, 330–334. [CrossRef]
37. Tewari, R.; Thyagarajan, K. Analysis of tunable single-mode fiber directional couplers using simple and accurate relations. *J. Lightwave Technol.* **1986**, *4*, 386–390. [CrossRef]
38. Pu, S.; Luo, L.; Tang, J.; Mao, L.; Zeng, X. Ultrasensitive Refractive-Index Sensors Based on Tapered Fiber Coupler With Sagnac Loop. *IEEE Photonics Technol. Lett.* **2016**, *28*, 1073–1076. [CrossRef]
39. Ranjbar, Z.; Jannesari, A.; Rastegar, S.; Montazeri, S. Study of the influence of nano-silica particles on the curing reactions of acrylic-melamine clear-coats. *Prog. Org. Coat.* **2009**, *66*, 372–376. [CrossRef]
40. Payne, F.P.; Hussey, C.D.; Yataki, M.S. Polarisation analysis of strongly fused and weakly fused tapered couplers. *Electron. Lett.* **1985**, *21*, 561–563. [CrossRef]
41. Bai, Y.; Miao, Y.; Zhang, H.; Yao, J. Simultaneous Measurement of Temperature and Relative Humidity Based on a Microfiber Sagnac Loop and MoS$_2$. *J. Lightwave Technol.* **2020**, *38*, 840–845. [CrossRef]
42. Han, C.; Ding, H.; Li, X.; Dong, S. Temperature insensitive refractive index sensor based on single-mode micro-fiber Sagnac loop interferometer. *Appl. Phys. Lett.* **2014**, *104*, 181906.
43. Vargas-Rodriguez, E.; Guzman-Chavez, A.D.; Garcia-Ramirez, M.A. Tunable optical filter based on two thermal sensitive layers. *IEEE Photonics Technol. Lett.* **2018**, *30*, 1776–1779. [CrossRef]

MDPI

St. Alban-Anlage 66

4052 Basel

Switzerland

Tel. +41 61 683 77 34

Fax +41 61 302 89 18

www.mdpi.com

Sensors Editorial Office

E-mail: sensors@mdpi.com

www.mdpi.com/journal/sensors